Queen B

Biology, Rearing and Breeding

by

David R Woodward

B.Sc., M.Sc.(Hons), Ph.D.

NORTHERN BEE BOOKS

To the three little workers

Lucinda, Jacqueline and Angela

...from the big drone

This volume was originally published in Balclutha, New Zealand as a supporting text to Queen Rearing courses developed by the author at Telford Rural Polytechnic

First published 2007 ISBN 978-0-473-11933-1

Reprinted in 2009 by Northern Bee Books, Scout Bottom Farm, Mytholmroyd, Hebden Bridge (UK)

This edition 2010

ISBN 978-1-904846-35-2

Printed by Charlesworth UK, Wakefield

Table of Contents

Chapter One: Queen Bee Biology

1.0 INTRODUCTION

The queen bee plays a pivotal role in the survival of the colony. Born from a fertilised egg and fed royal jelly throughout her larval life, she has a life expectancy of one to three years.

The queen is the mother of all bees in the hive. Her daughters, the workers, carry fifty percent of her genetic inheritance, while her sons, the drones, receive all their genetic information from the queen.

The queen is promiscuous. On the wing she may mate with up to seventeen drones over two to four days. Drones pay the ultimate price for an opportunity to mate with the queen, dying a violent death as their penis snaps in half following ejaculation!

The queen is the workers' 'surrogate father', as she carries their fathers' genetic material (spermatozoa) in an abdominal receptacle, the spermatheca, and uses this sperm to fertilise eggs that later develop into new queens or workers.

The queen is not the ruler of the hive. Her behaviour is strongly influenced by the workers. Her activity in the hive is partly dependant upon the stimuli that workers receive outside the hive. This stimuli is from plants that provide the all-essential food rewards of pollen and nectar needed to rear the young, and for the hive's survival over winter.

The queen is an egg laying machine, unable to look after the young that she produces. She may lay up to fifteen hundred eggs per day at the peak of her production and as many as a million eggs in her life-time.

A retinue of workers attend to the queen; these 'ladies-in-waiting' groom and feed the queen, providing for her every need so that she may carry out her important reproductive role.

The food that the queen receives is converted into egg development in her two large ovaries that provide an almost continuous supply of eggs to maintain the fifty thousand strong hive population.

Apart from egg laying, the queen's other important role is the production of queen substance, pheromones that are licked off her body and passed, via the workers, throughout the hive. Queen substance prevents the production of new queen bees, inhibits worker egg laying, stimulates worker foraging, and maintains cohesion in the hive.

As the queen ages the production of queen substance declines and the ability to inhibit the production of new queens subsides. Workers are stimulated to produce new queens to ensure survival of the hive. Without a queen, production of new workers ceases, the hive population declines and the remaining workers are unable to defend the hive from marauding robber bees from other hives.

The aging queen may initially attempt to destroy any new queen bees that develop, before they emerge from their cocoons. Eventually the hive population becomes too great for the queen to control solely on her own. She must depart the hive with a swarm before a new queen emerges, carrying with her sufficient bees to establish a new colony in a far off destination. The young workers assist her by transporting sufficient honey stores in their stomach, from the hive, to ensure a new colony can successfully establish itself in a new location.

Failure to leave the hive with the swarm during favourable weather could result in the aging queen facing the ultimate penalty, a fight to the death with a young, newly emerged queen. The victor, usually the young queen, ensures no further competition by destroying all other developing queens in their cocoons before they hatch.

Eventually, the young victorious queen departs the hive on her mating flights, returning with the remains of the last drone's penis with which she mated, protruding from her abdomen. After a few days she begins her newly ordained role of egg laying and pheromone production.

And so the queen has a demanding and vital role. She is the one bee among the fifty thousand that lays fertilised eggs. She has a perilous task to ensure survival of the hive through reproduction and colony cohesion in order that her genes, together with her deceased male partners' genetic information, may be passed on to future generations.

1.1 CASTES OF THE HONEY BEE

As part of the development of a complex social system, bees have evolved in a specialised manner. The honey bee colony is actually a super-family of individuals that are related to each other and depend on each other for the survival of the colony.

The colony is made up of one mother (the queen), thousands (50-60,000) of her daughters (the workers) and hundreds of her sons (the drones).

The queen, the workers and the drones are not able to survive on their own for any length of time. They each have particular functions within the colony that they perform and have specialised organs that help them to do this. These specialised bees are known as castes and each has an important role to play.

For an explanation of terms used in this book refer to the Glossary.

a) Queen

The major role of the queen (figure 1.1(a)) is the production of eggs and queen substance, pheromones that affect the behaviour of the bees in the hive. The queen does not control all activities that occur but she does have a significant influence on the ultimate survival of the colony. In a beehive there is normally only one queen. On rare occasions two queens may co-exist for many months, particularly in a mother-daughter relationship, however this rarely occurs when two virgin queens emerge at the same time or when the queens are unrelated (such as when a beekeeper introduces another queen). When virgin or unrelated queens are placed together there is usually a fight to the death.

The colony may raise more than one queen at a time. If this happens, the first queen to emerge from her queen cell (cocoon) seeks out other queen cells and destroys them. Queen cells hang vertically in the hive and are much larger than worker and drone cells. Queen cells are either found on the face of the comb (if the hive becomes queenless or the queen is failing), or in large numbers along the bottom of the comb (if the hive is preparing to swarm).

When a queen cell is sealed it has a tapered capping with a blunt tip. When the queen is ready to emerge the workers remove wax and fibres from around the tip of the cocoon (often referred to as crowning) to assist queen emergence (see figure 2.2.12(c)).

Figure 1.1(a) Queen bee (with blue disc) surrounded by workers.

Figure 1.1(b) Drones.

Five to six days after she emerges, the virgin queen leaves the hive on her mating flight. While in the air, she will mate with an average of ten but as many as seventeen drones, in one to five flights over a period of two to four days. Two to three mating flights per day are normal. Drones are attracted to the queen by pheromones that she emits. After mating the sperm from each drone is stored inside the queen in a sac called the spermatheca which holds from 5.3 to 5.7 million sperm. Other than during swarming, the mating flights are the only time the queen will leave the hive.

Within two to three days of her mating flight, the queen starts to lay eggs. The queen moves over the comb, examining each worker cell (hexagonal cavity) with her antennae and forelegs before laying an egg into it. Most cells in the comb are worker cells, but some are larger drone cells. The queen normally lays fertilised eggs in worker cells and unfertilised eggs in the larger drone cells. When she first begins to lay eggs (after mating), the queen often produces eggs faster than she can find cells to lay in. In such cases the workers may eat the surplus eggs.

Two to four years after mating the queen will run out of sperm and an increasing amount of drone brood can be seen in the comb. At this stage the old queen is usually superseded and killed by the workers. Thus the amount of sperm stored in the spermatheca determines the life span of the queen which is on average less than four years.

b) Drones

The only known function of the drone (figure 1.1(b)) is to mate with a virgin or partially mated queen. They do not collect food, sting, produce wax, or feed the young. They either beg for food from workers or rob it from cells.

Drones are reared in larger hexagonal cells than workers (see figure 2.3.2), on a 9-13° angle from horizontal. The drone cells have a markedly convex capping. Drone brood is normally found around the outside of worker brood or along the bottom of the comb.

Drones become sexually mature about twelve days after emerging and on most warm afternoons will fly out to find a queen on her mating flight. After mating, a drone dies within about one hour, as the reproductive organs are torn from its body. Drones that do not mate live, on average, 21-32 days during spring to mid-summer, but from late summer to autumn drones can survive for up to 90 days. Workers stop feeding the drones as the food supply decreases and expel them from the hive in late autumn, so drones are not usually present in the hive during winter. The population of drones in a

Figure 1.1(c) Workers.

hive increases during spring (from a few hundred), peaks in the summer (at several thousand), and decreases to low levels during the autumn.

c) Workers

Workers (figure 1.1(c)) make up the majority of the bees in the beehive, which can reach a size of 50-60,000 bees or more in summer. Workers are females, but do not usually lay eggs. They carry out a wide range of duties inside and outside the hive; this is dictated by their age and physiological maturity of internal organs. The workers duties in the hive include cleaning out cells, feeding larvae, grooming and feeding the queen, producing wax and building honeycomb, receiving nectar from forager bees, converting nectar into honey and fanning and guarding the hive. A foraging bee collects nectar, pollen, propolis, honeydew and water according to the hive's needs.

The worker brood (eggs, larvae and pupae) is found in hexagonal worker cells constructed at a 9-13° angle to the horizontal. The worker cell has a slightly convex opaque wax capping. The capping varies considerably in colour, from light yellow to dark brown depending on the age of the wax comb.

During summer workers live for 15-38 days. Workers reared during spring and autumn live for 30-60 days. The 'winter bees' live on average about 140 days because they have a higher body protein level and food has been stored during summer, so the requirements for foraging are minimal. Long-lived winter bees maintain colony survival throughout the winter and enable brood to be reared and incubated early in spring without the initial need to forage for food.

1.2 ANATOMY AND LIFECYCLE OF THE HONEY BEE

1.2.1 General insect anatomy

Honey bees belong to the class Insecta - the insects - so they have the following adult insect features:

     (1)   Three main body sections; head, thorax, abdomen
     (2)   Three pairs of legs attached to the thorax, or middle section

Insects have one or two pairs of wings; the honey bee has two pairs of wings that beat together, and one pair of antennae (feelers). Spiders and mites are not classified as insects because they have two body sections and eight legs.

1.2.2 Complete and incomplete lifecycle

Honey bees have a complete lifecycle (complete metamorphosis), that is, they develop through four stages: egg, larva, pupa and adult (figure 1.2.2a). This compares

Figure 1.2.2(a) Complete lifecycle of honey bee (complete metamorphosis).

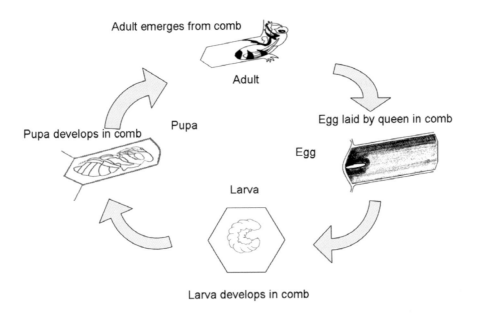

Figure 1.2.2(b) External anatomy of the queen honey bee and internal glands (Adapted from *Anatomy of the Honey bee*, by R.E. Snodgrass. Copyright © 1956 by Cornell University; also excerpted from The New Complete Guide to Beekeeping, © 1994 by Roger A. Morse. Reprinted with permission of the publisher, The Countryman Press/ W.W. Norton & Company, Inc.).

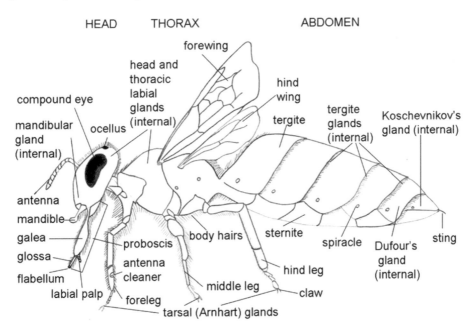

with insects such as aphids that have an incomplete lifecycle and have three stages: egg, nymph, and adult.

(i) Egg

The egg (or ovum), develops in a tubule called the ovariole, inside one of two ovaries of the queen bee.   When the egg is laid, sperm may be released from the spermatheca onto the egg. The egg has an outer shell called the chorion, with tiny perforations at one end called the micropyle. The micropyle allows sperm to enter the egg for fertilisation to occur. If the egg is fertilised, it will develop into a female bee (worker or queen). If sperm is not released by the queen, and the egg is not fertilised, it will develop by parthenogenesis into a drone. As the queen lays an egg she glues it to the bottom of the cell at one end so the egg stands upright. The egg is 1.3-1.8 mm long, weighing 0.12-0.22mg and is white and sausage shaped. Over three days the egg gradually keels over and the egg shell gradually dissolves. The dissolving of the shell appears to be unique to honey bees.

(ii) Larva

When the embryo inside the egg is fully formed the egg develops into a larva. The larva, or grub, is white and grows very rapidly, developing through five stages or 'instars' and moulting each time. The outer skeleton is shed as the larva outgrows it. The larva has no eyes, legs, wings or sting but has a well-developed mouth as most of the time is spent eating! The internal organs are made up of a well-developed stomach (or ventriculus), and excretory (malpighian) tubules that are cut off from the hind gut so that the stomach contents are only voided after the cell is capped. At this stage feeding is finished. The larva now stretches out along the cell and spins a silk cocoon with the aid of the spinneret and silk gland. This is the prepupal stage. The larva then moults and pupates. The final larva weight of a worker (140mg), queen (250mg) and drone (346mg) is 900, 1,700 and 2,300 times the egg weight, respectively.

(iii) Pupa

While in the cocoon the larva radically changes its form. This is the pupal stage and the change in form from the larva to the adult is called metamorphosis. Antennae, proboscis (tongue), compound eyes, wings and legs develop, and the head, thorax and abdomen become defined. The eyes change in colour from white, to pink, to purple, to black. When the adult is fully developed it moults for a sixth and final time and chews its way out of the tip of the cocoon.

(iv) Adult

The newly emerged adult (or callow), is white and fluffy. Over the next 24 hours the cuticle (exoskeleton) hardens, providing protection and reducing water loss, as well as providing an attachment for muscles. A dense coating of branched hairs, that are important in gathering pollen, cover the exoskeleton - especially on the thorax of workers.

1) Head

The head is well developed and contains the two compound eyes (figure 1.2.2(b)), made up of many individual hexagonal facets, or ommatidia. On the top of the head are three ocelli or simple eyes probably used for detecting light intensity. Connected to the front of the head are two antennae. Each antenna is made up of a base, or scape, and a series of segments that make up the flagellum. The flagellum acts like a nose, containing many sensory hairs that are important in detection of pheromones and for sending nerve impulses to the brain.

The mouthparts are made up of the mandibles (jaws) and the proboscis (tongue). The mandibular glands discharge royal jelly into the base of the mandibles in workers and may also help to soften wax. In the queen, the mandibular glands are involved in the production of queen substance.

The proboscis is made up of a number of parts and can be tucked under the head when not in use. When the proboscis is in use, the various parts form a tube and close over the mouth so that nectar, or dilute honey, can be sucked up, just like sucking liquid through a straw. A sucking pump that is present at the top of the mouth provides suction. The tube of the tongue is made up of the two galea in front and two labial palps behind. Inside this tube is the glossa and at the tip is the flabellum. The salivary glands are made up of two glands: the head and thoracic labial glands, which discharge into the top of the proboscis to assist with the digestion of food. The brood-food, or hypopharyngeal glands, discharge into the mouth and are important in the production of royal jelly in young adult workers and then, in older adults, produce enzymes, such as invertase, used for inverting nectar sugars into honey sugars. The head is joined to the next body section, the thorax, by a slender, flexible neck.

2) Thorax

The thorax contains three segments. The first segment is the prothorax, the second the mesothorax, the third the metathorax and, what appears to be the fourth segment of the thorax, is actually the first segment of the abdomen. The thorax has three pairs of spiracles (breathing holes) and connected to the thorax are three pairs of legs (fore, middle and hind legs) and two pairs of wings. The thorax is involved with locomotion and contains muscles for driving the wings and legs. Each leg is made up of a number of parts the coxa, trochanter, femur, tibia and tarsus. The coxa is connected to the thorax, while at the other end of the leg is the tarsus. The tarsus is made up of a swollen basitarsus and several tarsal segments and at the tip are the claw and the arolium. The claw is used for holding onto rough surfaces, and the arolium - a suction pad - allows bees to walk on vertical smooth surfaces.

On the front pair of legs at the join between the tibia and the tarsus is a notch called the antenna cleaner. The bee drags its antenna through this notch to clean off pollen.

On the outside of the tibia on the hind leg of worker bees, is the pollen basket, or corbicula. The corbicula is made up of a collection of long hairs used for storing pollen and propolis. On the inside of the basitarsus of the hind legs are rows of hairs called scopae and at the joint between the tarsus and basitarsus is the rake, or rastellum, and the pollen press, or auricle. The worker rubs the two inside rows of combs together and the rake deposits pollen on the pollen press. By articulation of this joint pollen is forced up into the corbicula.

The bee has four wings attached to the meso and metathorax, a larger pair of forewings and a smaller pair of hind wings. The wings are transparent and strengthened by veins. The two pairs of wings beat together in flight at a rate of over 200 cycles per second and to keep them working in unison a row of hooks or hamuli on the leading edge of the hind wing couple together with a curved fold on the trailing edge of the forewing.

The thorax is connected via the constricted waist to the abdomen.

3) Abdomen
The abdomen is made up of a series of overlapping plates of exoskeleton; on the upper abdomen these are known as the tergites and on the underside of the abdomen they are called sternites. The abdomen has ten segments but the first segment forms the rear of the thorax and the last three segments are reduced and concealed in the seventh segment. Thus the abdomen appears to have six segments. On the sides of the abdominal segments of the tergites are seven pairs of spiracles. These are the breathing holes for the bee, with three further pairs on the thorax. On the ventral (under) side of the abdomen in workers, there are four pairs of wax glands that secrete liquid wax into the wax mirrors on the external surface. The wax hardens into scales on the wax mirrors and each scale is removed by the hind legs, passed to the

mandibles, and with the help of the forelegs, manipulated into honeycomb. Near the tip of the abdomen, behind the seventh abdominal, dorsal plate, is the Nasonov or scent gland. Workers expose this gland when 'scenting', for attracting foraging workers from the hive and the queen on her mating flight back to the hive entrance. At the tip of the abdomen in the worker and queen is the sting. The sting shaft is made up of a stylet above and two barbed lancets below with a poison canal running through the middle.

The abdomen also contains the expanded honey stomach or crop (in workers) and digestive tract, tracheal air sacs, dorsal blood vessel, ventral nerve chord, and reproductive organs.

1.2.3 Anatomy of the castes
a) Queen

The major functions of reproduction and pheromone production are reflected in the body structure of the queen. For reproduction, the queen bee has a longer abdomen than a worker bee, especially after she has mated and begins egg laying. The expanded abdomen houses a well-developed pair of ovaries, which fill most of the abdominal cavity, and the spermatheca, the sac for storing semen. Each ovary contains 150-180 thin-tubed ovarioles.

The queen has several glands involved with pheromone production (figure 1.2.2(b)). The mandibular glands are well developed for queen substance production. On abdominal tergites 4 to 6, the queen has tergite glands producing pheromones that operate in a similar way to queen substance and that also attract drones during mating. The Koschevnikov gland, located at the base of the sting, produces pheromones to attract workers. The gland degenerates in one-year-old queens. The tarsal (Arnhart) glands secrete footprint pheromones from the arolium which, together with the mandibular gland, inhibits queen cell construction by the workers. Other non-pheromone producing glands include the head and thoracic labial glands that are well developed for processing food. Dufour's gland discharges its contents into the sting chamber. The function is unknown but may be involved in poison secretion or production of a waxy covering for the egg. The wax glands are absent in the queen.

To defend herself, at the tip of the abdomen, the queen has a curved sting with very small barbs on the lancets. The sting of the queen is used to kill other queens that may emerge in the hive. The queen does not die if she stings another queen as she can withdraw her sting from her victim but, if stung, may die soon afterwards. The compound eye is small with an average of 3,900 facets in each eye and used mainly for navigation on mating flights. The mandibles are very sturdy in the queen, possibly for grabbing and chewing other queens and for cutting small holes in the sides of other queen cells before stinging them. The proboscis of the queen is shorter than the worker. The queen bee has a bald dorsal thorax and the corbiculae (pollen baskets) on the hind legs are absent.

b) Drones

The one major function of the drone is to mate with a queen and the drone has many adaptations for this purpose. Drones are larger in size than workers and wider than the queen. Drones have a well-developed pair of testes and mucous glands as well as the penis or endophallus. They have large compound eyes that meet at the top of their head. The compound eye, with an average of 8,000 facets, is used for locating queen bees on the wing. The drones have large wings and are strong fliers, as they need to fly for the queen as well as themselves during mating, and have a rounded abdomen without a sting. The drone has one extra flagellum segment on the antennae possibly for detecting queen pheromones. Drones have a short tongue, small mandibles and no wax glands or corbiculae. Most of the glands present in the queen and workers are absent from the drone.

c) Workers

During the first 21 days after emergence, the worker carries out house duties inside the hive and after this period, foraging duties outside the hive. The adaptations for these tasks are reflected in her anatomy. Workers are smaller in size than the queen and drone. The worker bee compound eye, with an average of 6,300 facets, is used for locating food sources and for navigation. The mandibles are less developed than the queen but still sturdy and are used for eating, collecting pollen, manipulation of wax, and general house cleaning. The worker has a longer tongue than the queen and drone, for sucking nectar from flowers.

The worker has a dense covering of body hair, with long hairs on the thorax for gathering pollen. The pollen press and pollen combs are used for collection and transfer of pollen to the corbiculae on the hind legs, where the pollen is stored during transport. The worker has a straight sting; the lancets of the sting are highly barbed so they cannot be withdrawn from their victim. If the worker stings it dies soon after, when the internal organs are torn from its body. The workers have rudimentary ovaries with 2-12 ovarioles per ovary and rarely lay eggs. Because workers do not mate the eggs laid are not fertilised and the eggs develop into drones. The hypopharyngeal (brood-food) and mandibular glands are well developed for royal jelly production. The worker has eight (four pairs) wax glands on the ventral abdomen. The salivary labial glands are well developed for converting nectar sugars into honey sugars and the Nasonov gland is well developed to assist with orientation of workers, especially at the hive entrance.

1.3 REPRODUCTIVE SYSTEM OF THE DRONE AND QUEEN

Reproductive organs are well developed in the queen and drone, with rudimentary reproductive organs present in the worker.

i) Queen Reproductive Organs

The female reproductive cells (called ova or eggs) develop in the ovarioles inside the ovaries. The queen has two large pear-shaped ovaries (figure 1.3(a)) that are made up of closely packed tubules called ovarioles. Each ovary has 150-180 ovarioles. Each egg has 16 chromosomes and is genetically different from every other egg produced by the queen. As the eggs mature they move down the ovarioles and are supplied with nutriment by nurse cells that are absorbed by the egg as it develops. The fully developed eggs are discharged first into the paired (lateral) oviduct, second the common (median) oviduct, third the vaginal chamber, fourth the vaginal passage, and finally out through the vaginal orifice.

As eggs pass through the median oviduct, the valve fold presses each egg against the spermathecal duct, where sperm, released from the spermatheca (the sperm storage sac), make contact with and eventually fertilise the egg.

ii) Drone Reproductive Organs

The male reproductive cells (called spermatozoa or sperm) are produced in the testes. Each sperm produced contains 16 chromosomes and is identical to every other sperm produced by a single drone. The testes are a pair of flat structures in the abdomen of the drone. From each testis a coiled duct called the vas deferens extends into an enlarged sac called the seminal vesicle (figure 1.3(b)).

The two seminal vesicles enter the lower ends of a pair of large swollen mucous glands which discharge into a single tube, the ejaculatory duct. This duct opens into the penis. The penis is made up of a large bulb that contains a mass of viable sperm, a narrow neck (cervix) which breaks away after copulation, a fringed lobe, called the fimbriate lobe, and a pair of protruding horns called the bursal cornua. The penis discharges through an external opening. As the drones become sexually mature, spermatozoa produced in the testes migrate down the vas deferens into the seminal vesicles where they accumulate until mating.

Figure 1.3(a) Abdominal section showing reproductive organs of a queen (Redrawn from Ruttner, F. 1976. The instrumental insemination of the queen bee).

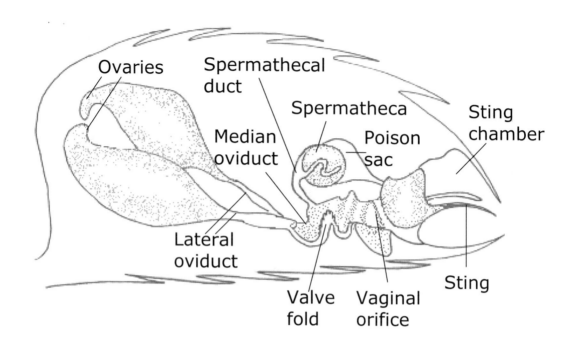

Figure 1.3(b) Drone reproductive organs (internal) (Redrawn from Snodgrass R.E. and Erickson, E.H. 1992. *In* The Hive and the Honey Bee. Dadant & Sons, Inc.)

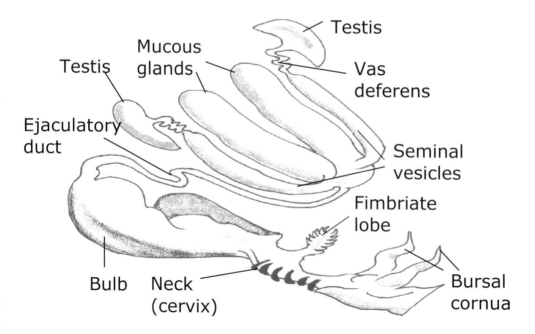

1.4 CASTE DETERMINATION

Queen bees or workers can develop from the same type of egg. Eggs that are not fertilised with sperm by the queen develop into drones. Fertilised eggs may develop into either queen bees or workers depending on the quality and quantity of food fed during the larval stage.

The queen bee determines whether the egg is to be laid in a worker or a drone cell by inspecting the cell. She does this with her forelegs or the angle of her abdomen during the egg laying process. If the egg is going to be laid in a drone cell, the queen bee will not release spermatozoa from the spermatheca.

Under certain conditions, usually when the hive is without a queen, workers can lay eggs. As workers never mate, the eggs they lay are unfertilised and develop into drones only.

Eggs laid by workers are smaller than eggs laid by queens and are placed on the side of the cell rather than at the bottom of the cell as the worker abdomen is too short to reach the base of the cell when inserted for egg laying.

The egg takes three days to develop. The larva that develops from the egg is fed by nurse bees on either 'royal jelly' or 'worker jelly', depending on the type of cell the larvae develops in. Royal jelly and worker jelly consist of secretions of the hypopharyngeal (brood-food) glands and mandibular glands in the head of worker nurse bees.

During the first 24 to 36 hours of larval development workers and queen bees are fed similar amounts of food. During the second day the diet for larvae that will become workers or queens differs. Worker larvae receive a light feeding of worker jelly for the first three days of development, produced from secretions of the mandibular and hypopharyngeal glands of nurse bees (figure 1.4). Mandibular gland secretions are white and made up of lipids (fats) whereas hypopharyngeal gland secretions are clear and made up of protein. Worker jelly consists of 12% sugar, with the predominant sugar being glucose. During the fourth and fifth days, worker larvae receive light feeding of worker jelly made up of secretions from the hypopharyngeal glands only. Honey and pollen are added to the diet increasing the sugar content of the jelly to 47% and the predominant sugar changes to fructose. By contrast, larvae developing in queen cells are fed royal jelly in copious amounts. During the first three days the royal jelly fed to these larvae is produced from nurse bee secretions of the mandibular glands only. The mandibular gland secretions in royal jelly contain 18 times higher

Figure 1.4 Factors determining development of the worker, queen and drone.

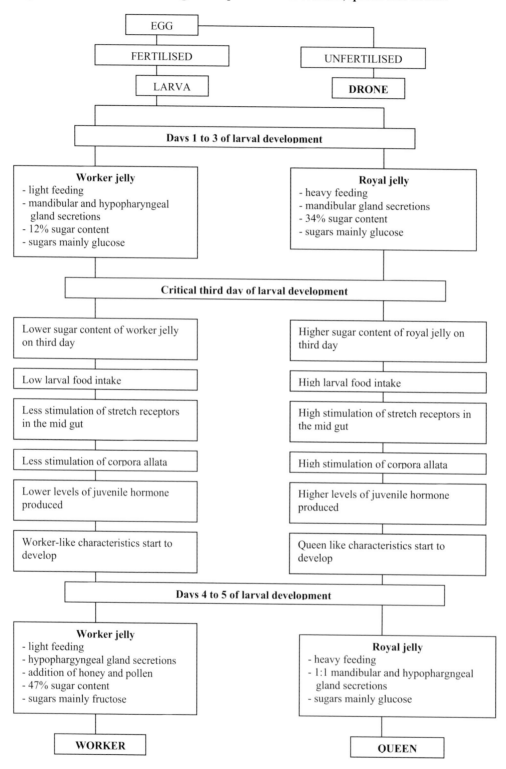

levels of biopterin and ten times more pantothenic acid than worker jelly. The royal jelly consists of 34% sugar and is predominantly glucose. During the fourth and fifth day larvae in queen cells continue to receive copious royal jelly made up of 1:1 mandibular and hypopharyngeal gland secretions with glucose remaining the predominant sugar.

The quality and quantity of jelly fed to larvae on the third day is critical in determining development of worker or queen characteristics. Research, although not conclusive, suggests the following possible pathway for female caste determination. Larvae feeding on royal jelly with 34% sugar content, compared to 12% in worker jelly, have a higher food intake. The higher food intake provides a greater stimulation of stretch receptors in the midgut and this provides greater stimulation of the corpora allata, a large globular organ found on the sides of the oesophagus. Greater stimulation of the corpora allata results in higher levels of juvenile hormone produced and this in turn results in queen-like characteristics developing. Worker larvae, feeding on worker jelly with lower sugar content, have a lower food intake on the third day so there is less stimulation of the stretch receptors and corpora allata. This results in less juvenile hormone produced, and hence worker-like characteristics developing.

The nurse bees feed the queen bee larva royal jelly for an average of 4½ days before the queen cell is sealed over with wax and feeding ceases.

1.5 CASTE DEVELOPMENTAL STAGES

The queen bee develops through the egg stage in three days and then hatches into a larva. The queen bee larva remains in the unsealed stage, on average, for only 4½ days (range 3 - 5 days). Once the cell is sealed the larva spins a cocoon and develops into a prepupa and then pupates. The sealed stage lasts for 8½ days. The queen bee emerges as an adult after 16 days (figure 1.5). Worker development takes three days as an egg, the unsealed larval stage lasts 5½ days and the sealed larva, prepupa and pupal stage lasts 12½ days, with the adult emerging after 21 days. Drone eggs develop in three days, spending 6½ days as an unsealed larva, and 14½ days as a sealed larva, prepupa and pupa, emerging as an adult after 24 days. The reduced time for queen development ensures that a colony can quickly replace a queen that has been lost. Development times from egg to adult may vary considerably depending on temperature and nutrition. The workers normally maintain the brood temperature at 35°C; a lower brood temperature will delay development.

Figure 1.5 Developmental stages of the three castes of the honey bee.

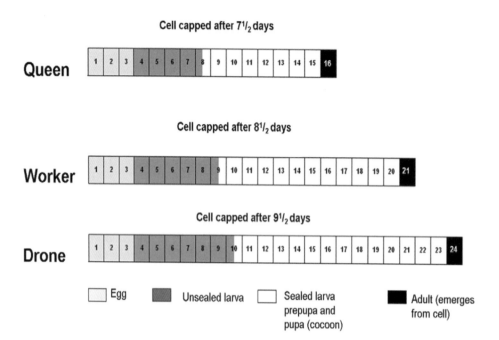

1.6 DRONE PRODUCTION AND DEVELOPMENT

There are a number of factors that determine drone production in a hive. The strain of bee influences drone numbers, with Italian queens producing more drones earlier in the season than darker strains such as Caucasian or Carniolans.

Weaker hives, and hives infected with diseases such as chalkbrood, produce fewer drones than disease-free hives.

Where a virgin queen bee fails to mate or is infected by a virus she may become a drone laying queen, and lay only unfertilised eggs in worker cells. Sometimes if a queen is partially mated, she will lay fertilised eggs for a period of time and then lay unfertilised eggs that develop into drones.

Drone larvae are fed much more jelly than worker larvae because of their larger size, but also the drone diet has a wider range of proteins. The composition of the diet of older drone larvae changes as workers add more pollen and nectar.

Adult drones are fed entirely by workers for the first few days of their lives and then gradually begin feeding themselves from honey cells. Young adult drones from one to eight days old are fed a mixture of brood-food secretions from nurse bees, pollen and honey. Once the drones reach about eight days old they feed themselves mainly on honey. The honey provides energy for mating flights.

Drones reach sexual maturity about 12 days after emergence. At this stage sperm production is complete and mature sperm are stored in the seminal vesicles until mating.

A good supply of pollen, nectar and nurse bees are required in a colony to rear and maintain adequate numbers of drones. These conditions usually occur from late spring until early autumn.

During autumn, workers stop feeding the drones and remove them from the hive. This means that usually drones are not present in the hive during the winter when conditions are unsuitable for mating and drones are a drain on food resources. The exception to this would be where the weather is milder over winter, and a hive has ample food stores, in which case it may then carry drones through the winter.

1.7 QUEEN CELL DEVELOPMENT

Natural queen cells develop in a hive under a queen rearing impulse (see 1.12). The queen develops from a fertilised diploid egg. The larva that develops from the egg is fed on copious amounts of royal jelly produced by worker nurse bees. The developing larva is not fed directly but floats on a bed of royal jelly, turning around in the queen cell as it feeds. Worker bees add wax to the queen cell to extend its length. The queen larva spins a cocoon on the fifth day of larval development. The cocoon is built around the sides and tip of the cell rather than completely surrounding it as occurs with workers and drones. A pupating queen is not completely protected from the sting of another queen due to the incomplete nature of the cocoon.

The queen larva stretches lengthways in the cell with the head down and develops into a pupa. At some point close to the time of queen emergence, wax is removed from the tip of the pupa by worker bees (crowning). Eventually the mature queen chews her way out of the tip of the cocoon using her mandibles.

1.8 QUEEN BEE POST EMERGENCE ACTIVITY

The queen emerges after 16 days, chewing out of the cocoon on her own. She feeds on honey, and possibly pollen, without assistance. She then learns to obtain food from workers. By the third day workers start to lick and groom her.

Virgin queen bees move about the brood searching for other virgin queens or queen cells. When virgin queens find each other they will fight. The two virgin queens attempt to sting each other by piercing the soft tissue of the abdomen between the inter-segmental membranes of the hard outer plates of the exoskeleton. The queens grasp each other by the legs in a fatal embrace from which emerges only one winner. Often a younger queen will successfully kill an older queen, but not always.

A virgin queen chews a hole in the side of other queen cells (usually above where the cocoon extends to) and then stings the developing queen bees inside. The queen will then begin to tear down the cell wall. Worker bees continue the cell's destruction and remove its contents. Nurse bees stop feeding other queen larvae once the first virgin queen emerges.

After three to five days, the virgin queen will take short orientation flights of only a few minutes in order to orientate herself to the area around the hive. The virgin queen has a short abdomen at this stage and becomes sexually mature within five to six days of emergence.

When the queen is ready to mate a court or retinue of worker bees assemble around the queen. The workers treat her roughly and the queen responds by piping.   Piping is a series of high pitched, pulsed sounds from the queen produced by operating her wing beating muscles without beating her wings. Piping causes workers to freeze on the comb until the queen stops piping. Workers encourage the queen to leave the hive and mate on warm afternoons and remain at the entrance releasing orientation pheromones from their Nasonov gland to assist the queen to locate the hive entrance.

1.9 QUEEN AND DRONE MATING

The rough treatment of the queen by the workers may induce mating flights. The queen will mate an average of ten but up to seventeen times during one to five flights over a period of two to four days, depending on the weather, and store the sperm in a sac called a spermatheca. In fine sunny weather, most virgin queen bees' mate within eight to nine days of emergence and all are usually mated after two weeks of good weather. The queen may take up to four weeks to mate if weather conditions are poor. After three to four weeks of confinement, a virgin queen often loses her ability to mate and will lay unfertilised drone eggs.

Mating flights normally last 5 – 30 minutes, at distances of up to several kilometres from the hive.

Mature drones fly for longer than workers and queen bees, with flights between 25 to 57 minutes, with an average of 31 minutes. Drones usually fly within 3 km of the apiary. They tend to congregate in specific locations called drone congregation areas, where they remain in flight waiting for virgin queen bees to arrive. There may be several such congregation areas within flight range of the apiary. Drones from many locations are attracted to these areas and the areas do not change from one season to the next even though used by a new generation of drones.

Mating normally occurs on sunny days at temperatures above 18^0C and wind speeds of less than 18 km per hour. Cloud cover, shade and direction of the colony entrance affect drone flight. Drones fly mainly in the afternoon (1400-1600 hours) with 3-4 flights on sunny days but with only one flight on cloudy days. The total time flying may be 2½-3 hours per day. They do not stop to rest during searching flights for a queen. There have been recordings of drones mating with queens from hives separated by up to 16 km but this would be rare.

Drones are attracted to the virgin queen by pheromones (chemicals released by the queen to attract the drone) and visually by movement of the queen. The virgin queen may fly up to a height of 7 – 17m with drones following the queen bee in flight. When a drone gets close to a queen the drone will grasp the queen bee's abdomen with all of its six legs. The abdomen of the drone curls downwards until it makes contact with the tip of the queen's abdomen (figure 1.9(a)).

If the queen bee is ready to mate she will open her sting chamber, the drone extends his penis and within a few seconds ejaculation occurs.

The penis consists of an endophallus that is everted during ejaculation. The end of the penis with the bursal cornua protruding (figure 1.9(b)) is inserted into the sting chamber of the queen. Spermatozoa are discharged from the seminal vesicles (figure 1.3(b)), together with mucous from the mucous glands, into the ejaculatory duct and accumulate in the bulb of the penis. When the penis is inserted into the queen's sting chamber, the drone is instantly paralysed, and releases the queen. The drone flips backwards and ejaculation occurs from the pressure of the drone's haemolymph as his abdomen contracts. White mucous, along with a pinky latticework of spermatozoa are discharged from the external opening of the penis into the vaginal orifice of the queen.

After ejaculation, the drone and queen bee separate and a popping sound may be heard as the penis snaps in half. The drone falls to the ground and dies within approximately an hour of mating. A portion of the penis remains in the sting chamber of the queen's abdomen. The remains of the penis can be clearly seen inserted into the rear of the queen and is known as the 'mating sign'.

Once the queen has mated, spermatozoa that have migrated into the common and paired oviduct are forced back into the vaginal chamber by muscle contractions of the queen. The valve fold helps to direct sperm into a canal called the spermathecal duct (see figure 1.10) and finally the spermatozoa reach the sperm sac, called the spermatheca. The migration of spermatozoa from the oviducts into the spermatheca is influenced by the temperature surrounding the queen for up to 40 hours following ejaculation or insemination, and the activity of the queen in the hive. Between 5.3 and 5.7 million spermatozoa are stored for the life of the queen bee (about 8 microlitres). Sperm stored in the spermatheca are supplied with nutrients from the spermathecal gland; this ensures the sperm remain viable for many years.

Each drone that mates with the queen bee dislodges the 'mating sign' of the previous drone. During successive mating, the sting chamber of the queen bee stays open and it is only on the last mating that the queen bee closes the chamber by cutting off the

Figure 1.9(a) Queen (left) and drone (right) mating on the wing (Redrawn from Gary, N.E. 1963. Observations of mating behaviour in the honey bee; and Gary, N.E. 1992. *In* The Hive and the Honey Bee. Dadant & Sons, Inc.)

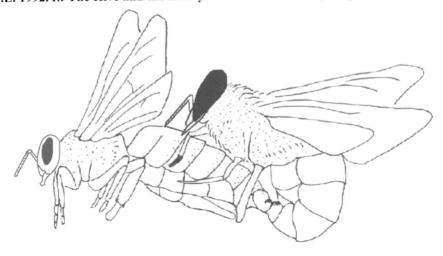

Figure 1.9(b) Drone reproductive organs (external) (Redrawn from Ruttner, F. 1976. The instrumental insemination of the queen bee).

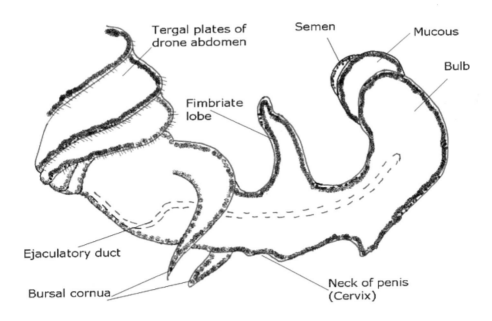

bulb of the penis, and returns to the hive with the mating sign protruding from her abdomen.

The 'mating sign' may therefore be removed by another drone mating with the queen bee, by the queen contracting her abdomen, or by being pulled out by the workers with their mandibles when the queen returns to the hive.

In some cases the queen bee may remove the 'mating sign' before she returns to the hive. This means that although a queen bee may not be showing a 'mating sign', she may still have successfully mated with many drones.

The queen will begin to lay eggs often after 12 hours but usually within two to three days of mating.

1.10 QUEEN EGG LAYING

As the queen's abdomen expands she is fed royal jelly high in protein. This is required for (i) the large demand for production of eggs and pheromones (ii) because the queen continues to grow after emergence and (iii) because the queen lives much longer than workers. The queen bee generally lives for one to three years. In unmanaged colonies, about 79% survived for one year, 26% of these queens survived for two years and none survived longer than three years. However, a queen may live as long as five years if she does not lay eggs at her full capacity.

About two to three days after mating, the queen will begin egg laying. Eggs move down the lateral and median oviduct into the vagina, and undergo cell division. One cell in the egg is absorbed and the remaining cell contains a nucleus with chromosomes. Each egg cell now contains 16 chromosomes. The eggs are now ready for fertilisation. The egg is pressed against the opening of the spermathecal duct by the valve fold (figure 1.10). A small 'S' shaped pump and valve, about half way along the duct, opens and allows a small amount of sperm and seminal fluid to travel from the spermatheca through the duct to fertilise the egg in the vagina. Presumably, when the queen lays eggs that develop into drones, the valve will close, semen will not be released through the spermathecal duct and the egg will not be fertilised.

Spermatozoa released onto the egg surface enter through a small pore, the micropyle, on the outer surface of the egg.

Figure 1.10 Spermatheca, duct and valve fold of queen (Redrawn from Dade, H. A. 1985. Anatomy and dissection of the honey bee. International Bee Research Association).

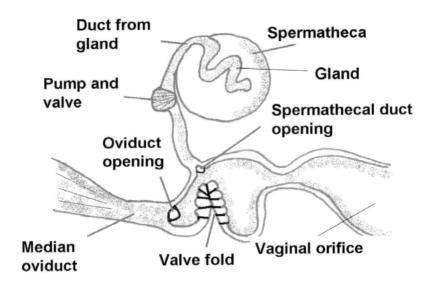

A fertilised egg is formed when the nuclei of the egg and one of the sperm unite. The fertilised egg now contains 32 chromosomes, 16 from the egg and 16 from the sperm and will develop into a female bee (either a worker or queen). An unfertilised egg has 16 chromosomes and will develop by parthenogenesis into a male bee (drone). Parthenogenesis is the development of an egg, without fertilisation, into a new individual.

The eggs are discharged from the vaginal orifice at the base of the sting into the bottom of a cell in the brood area of the hive.

The queen checks each cell with her forelegs and antennae to determine if it is of a suitable depth and diameter. She then pushes her abdomen into the cell and lays one egg on the base of the cell.

The queen bee lays up to 1500 eggs each day during the most active period of brood rearing.

1.11 QUEEN PHEROMONES

A pheromone is defined as a chemical message secreted externally resulting in a behavioural change in an animal of the same species. 'Pheromone' is a relatively new word, first suggested by German scientists in 1959. Pheromones have a major influence on bee behaviour. The use of pheromones or chemical messages is one of the major ways bees communicate and they are secreted from exocrine glands as a liquid and transmitted as a liquid or gas. Pheromones are detected by specialised cells called chemoreceptors either in the gaseous state (olfaction) or the liquid state (gustation). Pheromones are detected in very small quantities by honey bees.

When a virgin queen emerges she does not produce pheromones initially, this occurs slowly over the first three days. Once the queen reaches the third day of adult life workers are attracted to her by pheromones and start to lick and groom her.

The queen bee produces pheromones in the mandibular glands (e.g. 9–oxo–2–decenoic acid or 9ODA and 9-hydroxy–2–decenoic acid or 9HDA) and these pheromones operating in association with other pheromones are collectively referred to as 'queen substance'. Queen substance is groomed from the queen by the retinue of workers, and passed around the colony as workers exchange food and through body contact. Queen substance contains a number of chemicals not all of which have been isolated and identified, but it seems that some of them act as inhibitory substances. This means that when queen substance is produced by the queen bee and spread through the colony to the workers it inhibits the construction of queen cells by workers, and the growth of worker ovaries. Queen substance attracts drones to virgin queens on their mating flight, stimulates worker foraging behaviour and regulates movement and cohesion of swarms. It is the lack of this pheromone that causes the workers to start rearing another queen bee.

The tergite glands, on abdominal tergites 4 to 6, work in association with the mandibular glands producing pheromones that inhibit queen cell production, inhibit worker ovary development, act as queen recognition signals for workers, and stabilise her retinue. They also help to attract drones on mating flights and to induce copulation.

The Koschevnikov gland, located at the base of the sting, produces pheromones to attract workers. This may help to stimulate the feeding of the queen by nurse bees. The gland degenerates once the queen is a year old. From the combination of several pheromones the queen will have a retinue of attendant bees feeding and grooming her wherever she moves on the comb.

Virgin queens produce a pheromone that repels workers and other queens for the first two weeks after emergence. This pheromone is discharged as part of the faecal deposits from the rectum.

Workers produce pheromones from the Nasonov gland, one of which assists in the queen orientating to the hive entrance on her mating flight. Drones produce pheromones in the mandibular glands that are attractive to other drones flying in drone congregation areas.

1.12 QUEEN BEE REARING IMPULSES

Queen bees are produced by one of three queen bee rearing impulses in a hive: (1) Emergency (2) Supersedure (3) Swarming. Worker bees, guided by their recognition or lack of recognition of queen substance, will begin to rear a replacement queen under one of these three queen rearing impulses:

1.12.1 Emergency

This is the easiest impulse to understand. If the queen is removed from a colony, within a matter of minutes the workers are aware that the queen is missing. This is as a result of the bees that were attending the queen (feeding and grooming her) not being able to lick queen substance off the queen's body and pass it to the other workers in the colony. With the absence of queen substance, the workers are no longer inhibited from raising queen cells.

The workers will take several of the last fertilised eggs that the queen laid in worker cells and, when these eggs have hatched, begin to feed them copious amounts of royal jelly, to convert them into queens. Emergency cells are found in large numbers (5-20) on the face of the comb.

The key point to remember about this impulse is that it only occurs when the queen dies, is removed, or is isolated from the workers in the colony, so that there is no contact between the queen and workers.

1.12.2 Supersedure

This is the response by the workers to the signal that their queen is failing and needs to be replaced. She has not left the colony, so the emergency impulse does not operate. As the queen ages her egg laying rate and queen substance production decreases, the concentration of queen substance received by each worker declines and the workers are triggered to raise queen cells to produce a new queen that will replace the old queen. Since there is no emergency, the workers will make several (1-5) wax cups on the face of the comb and the old queen will lay eggs directly into them. When the old queen is failing or injured, it is quite usual to find her being replaced by a younger queen. Under these conditions, there may be a period of co-existence and it is not unusual to find a mother and daughter laying eggs in the same colony before the old queen dies.

1.12.3 Swarming

Rather than having a simple presence/absence of queen substance, swarming combines many factors. Overcrowding of the bees in the hive is a major factor. The age of a queen is also a factor. An old queen produces lower levels of queen substance and in a crowded colony this reduced amount has to be shared among many bees. Colony size also influences swarming. Large colonies in hives with ample volume are more likely to replace the queen by supersedure compared to smaller colonies in smaller volume hives that are more likely to replace the queen by swarming. The age of the workers may also be important with a higher proportion of young workers present when swarming occurs. The percentage of young workers remaining in the hive after swarming depends on the amount of brood in the hive.

A decline in concentration of queen substance per bee, together with an aging queen and overcrowding combined with other internal hive factors and external factors such as a light nectar flow and abundant pollen following a dearth in the hive, may cause the workers to begin queen cell (swarm cell) production. The aim is not only to raise a new queen but also the replacement of the colony.

The bees can confuse supersedure and swarming impulses. Queen cells begun simply to replace an ailing queen sometimes result in a swarm developing from the colony except that, unlike normal swarming, the new virgin queen leaves with the swarm.

Swarm cells are produced in large numbers (15-25) along the bottom and sides of the brood comb to ensure that a queen will emerge to take over the egg laying in the hive after the old queen departs. Most of the fertilised eggs in swarm cells are laid by the original queen but workers can and do move a small number of fertilised eggs or very young larvae from worker cells into queen swarm cells.

Swarming is the natural way the colony reproduces. About 60% of the workers and drones together with the old queen will leave the parent colony to start a new colony somewhere else.

In New Zealand swarming occurs from late September to December when the numbers in a colony are rapidly increasing. The swarming period for a single colony from initiating queen cell raising until the new queen has mated and starts egg laying takes about four to five weeks in good weather.

1.12.4 Colony reproduction by swarming

The first step in colony reproduction is the increase in colony numbers. As worker numbers increase in the hive, the queen starts to lay drone eggs in available drone comb. About three to four weeks after the queen has started laying drone eggs, worker bees start drawing down queen cells on the sides and lower edges of the combs. The queen will lay fertilised eggs in these queen cells. Workers will also transfer eggs and larvae into these queen cells and the resulting larvae will be fed continuously on royal jelly. Colonies normally swarm on the day the first queen cell is capped over or the day after, although the timing of swarming can vary considerably.

For up to ten days before swarming, the workers gorge themselves on honey, carrying about 36mg per bee in their honey stomach.

One week before swarming, workers feed less royal jelly to the queen and her egg laying decreases. The queen's ovaries shrink reducing the size of her abdomen, so that she will be light enough to fly with the swarm. The queen will be pushed, shaken about and treated roughly by the workers.   This action is to keep her moving.

Several days before swarming, a large number of bees will be seen 'resting' quietly on the bottom of the combs. On a fine, sunny, warm, calm day during the warmest part of the day, the 'scout bees' will perform a special dance that incites the bees to swarm.   The swarm emerges from the hive in a frenzy of flight. The number of bees leaving with the swarm will be 50-90% of the total colony population and is made up of bees of all ages, but up to 70% of workers less than ten days old will leave with the swarm. About 1% of the swarm population will be drones. The excited workers drive the old queen out of the hive. The first swarm to leave a hive is known as the prime swarm and has an average of 16,000 bees.

Within minutes of emerging, the swarm will cluster around the queen in a dense mass nearby e.g. on the branch of a tree. After the swarm has settled, scout bees will leave the swarm to search for a new nesting place. The returning scout bees perform 'wagtail' dances on the surface of the swarm cluster thereby indicating the distance and direction of the new nest site. The better the nesting place, the livelier is the scout's dance.

Before the swarm leaves from where it clustered, the bees clean themselves and start moving about. A loud humming noise can be heard. When the excitement reaches a climax, about ten bees will fly out of the cluster, hundreds follow and in a few seconds the entire swarm is airborne.

When the swarm arrives at the new nest site, scout bees alight at the entrance and release Nasonov pheromone to attract the swarm to the nest entrance. It takes only a few hours before the bees start new comb construction at their new site.   Within a few days a new brood cycle is started and a new colony is successfully established with the old queen.

Back at the old hive, a new queen bee will emerge from one of the queen cells about a week after the prime swarm leaves the hive.   An emerging virgin queen communicates her presence to the colony through producing pheromones and by piping. A piping queen inhibits workers from removing the wax and fibre around the tip of other queen cells (crowning), preventing further queen emergence.

The emerged queen then cuts a small hole in the side of the remaining queen cells and stings through the wall of the cells to kill the developing queens before they emerge. Workers assist in removing and dismembering the queens from the cells.

The virgin queen then leaves the hive to mate, returns and starts egg laying, so the original colony will continue to survive. Sometimes this virgin queen will leave with a second swarm two to four days after emergence and before mating. This is known as an 'after swarm' and contains an average of 11,500 bees and up to three virgin queens. The survival chance of these after swarms is very low, possibly because the virgin queen still has to mate and there are fewer workers to start a new colony. Whereas prime swarms, with mated queens, have a much higher chance of survival.

1.13 VISUAL PERCEPTION AND NAVIGATION

Bees have two compound eyes and three simple eyes called ocelli. The compound eyes are in fact a collection of individual facets. Each worker bee has approximately 6300 of these facets whereas the queen has only 3900 facets. The large number of facets enables the bee to distinguish very complicated shapes, forms and patterns, and to detect rapid movement - up to 200 movements per second. This is extremely beneficial for detecting flowers moving in the wind. The ocelli are thought to be used in detecting light intensity.

Bees use a combination of visual, olfactory (smell) and magnetic senses to locate food sources, mates, and nest sites.

Bees possess trichromatic vision, similar but not the same as humans. While humans can see the colours of the rainbow (red, orange, yellow, green, blue, indigo, violet), Karl von Frisch discovered that bees have a different colour spectrum from humans. Bees cannot see the colours at the long wavelength (700+ nanometres) end of the spectrum i.e. red. They do, however, see into the short wavelength (<400 nanometres) end of the spectrum, well below what humans can see (figure 1.13(a)). This enables bees to detect ultraviolet light. The order of sensitivity to colours for the honey bees is: ultraviolet > blue-violet > green > yellow > blue-green > orange. Bees can also see a colour called bee purple that results from the combination of ultraviolet and yellow.

Karl von Frisch realised there are primary and secondary methods used by honey bees for navigation and if the primary mechanism could not operate, secondary mechanisms could be used for navigating during flight.

The primary mechanism used for navigation is the compass position of the sun during the day and honey bees can compensate for the diurnal movement of the sun through the sky. Honey bees can calculate the movement of the sun through the sky during

Figure 1.13(a) The colour spectrum of honey bees and humans (scale nanometres) (Redrawn from von Frisch, K. 1955. The dancing bees).

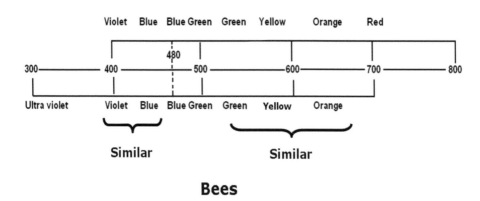

Figure 1.13(b) Bees cannot distinguish between the closed shapes in the top row, or between the open shapes in the middle row, but they can distinguish those in the top row from those in the middle row. The bottom row has two shapes that honey bees can easily distinguish between. The left shape is broken while the right shape is unbroken (Redrawn from von Frisch, K. 1955. The dancing bees).

the day even if they cease foraging for a while and then adjust their flight direction according to the sun's new position when they begin flying.

One of the secondary backup mechanisms for navigation is the ability of honey bees to determine the position of the sun on cloudy days. The ability to detect the ultra violet component of light from the sun allows bees to determine the position of the sun even when the sun is not visible.

Light from the sun is polarised so the direction of vibration of light waves changes in a regular pattern as the sun moves across the sky. These polarisation patterns can be detected by bees but not humans, through a patch of blue sky even with the sun obscured. It is the ultraviolet component of polarised light that is detected by workers.

On heavily overcast days when sun compass position and the ultra violet component from polarised light are not visible, another backup mechanism for navigation is used. Landmarks are used to locate food sources. Workers use landmarks to memorise the diurnal movement of the sun and use these landmarks in the sun's absence. Other secondary mechanisms include: orientation to the earth's magnetic field using detectors in the abdomen; floral odour plumes; and for workers and the queen on homeward flights, orientation to Nasonov pheromones released by workers at the hive entrance; as well as colony specific odours which allow discrimination between their colony and foreign colonies.

Queen bees returning to the hive after mating use certain cues to locate the hive entrance. Knowing which hive to return to is important. If the queen or workers enter the wrong hive by mistake they may be killed. Workers may be accepted into another hive under certain conditions e.g. during a honey flow, if they have nectar to offer which results in unbalanced apiaries and an overall reduction in honey crops. Bees are more likely to drift from a weak to a strong colony or from a queenless to a queenright colony. In the case of the queen, if she does not return to the hive after mating, the hive may die.

The homing ability of the honey bee develops over time as it learns the location of the hive in relation to surrounding objects. A bee's first flight is very short. The bee, on taking flight, turns to face the hive, hovers while it memorises the immediate surroundings and then re-enters the hive.   The distance of each flight is slowly increased until the bee becomes familiar with the entire foraging area around the hive. Karl von Frisch's explanation suggests that the bee's ability to find its way home is not the result of any inherited trait but is the result of the bee's ability to gradually memorise the position of the hive.

The honey bee may be able to determine the angle at which the entrance is located relative to other hives, and use cues such as silhouettes, stones and fence posts to distinguish its own colony entrance from all others. A bee cannot distinguish between shapes that look different to humans, for example between a dark square and dark triangle (figure 1.13(b) top line), but they can distinguish between a shape that is closed such as a dark square from one that is open such as the outline of a square (figure 1.13(b) middle line) but not between two closed or two open shapes. Also bees can distinguish between unbroken and broken shapes (figure 1.13(b) bottom line).

1.14 QUEEN AND DRONE ABNORMALITIES AND DISEASES

Sterile eggs: Infrequently, a newly mated queen begins and continues apparently normal egg laying, but most eggs do not hatch. The eggs eventually shrivel and are removed by worker bees. The cause is possibly genetic resulting from inbreeding and the queen should be replaced.

Scattered brood is usually due to the queen mating with drones that are closely related to her, resulting in the production of diploid drones that may occupy up to 50% of the brood. Under normal conditions worker bees eat diploid drone larvae within six hours of hatching from the egg, producing the characteristic spotted brood pattern.

Multiple egg laying in cells: When the queen first begins laying she may lay several eggs in one cell and sometimes eggs may be laid on the wall of the cell. Usually this will happen for a few days before the queen settles down to a regular pattern of egg laying. However, if older queens begin laying on the wall of the cell or chains of eggs or multiple eggs are laid in one cell or drone eggs are laid in worker cells, this may be a sign of Halfmoon syndrome. When workers lay eggs, because they are not fertilised they develop into drones and as their abdomen is shorter than the queen, they cannot reach the base of the cell to deposit the egg so one or more eggs may be laid on the side wall of the cell.

Drone laying: If a queen fails to mate, or mates with only a very limited number of drones, she may begin laying drone (unfertilised) eggs in worker brood. This may occur as soon as egg laying begins or some months after initial worker egg laying. This is due to a lack of sperm in the spermatheca of the queen, used to fertilise the eggs that develop into females.

Catalepsy is a shock or fainting of the queen due to a temporary nervous disorder as a result of a beekeeper picking up the queen by her wings. This does not happen regularly and queens can be handled by their wings with care.

Abnormalities: Defects such as queen bees with abdomens curved to one side, stunted wings, loss of a leg or part of a leg or an antenna, result in poor performing queens.

Black queen cell virus (BQCV) affects queen bee pupae forming a black ring around the tip of the queen cell. The queen pupa dies and turns black prior to emergence.

Nosema (*Nosema apis*): This disease is caused by a single celled protozoan parasite which invades and destroys the cells lining the midgut (ventriculus) of adult honey bees. Nosema infects workers, drones and queens. Although a beekeeper cannot recognise an infected queen, the reduction in the adult population of a colony in late winter and early spring is a common symptom. When a queen ingests Nosema spores the queen will usually be lost within 10-15 days. The last eggs laid by an infected queen often shrivel and fail to hatch making it difficult for the worker bees to rear a replacement queen. Queen bees may become infected by contact with infected workers in mating nuclei, mailing cages or in package bees. The brood-food glands of infected workers do not fully develop so there is a reduction in royal jelly production.

Halfmoon Syndrome (Disorder): This is not a disease but is thought to be due to abnormalities of the queen bee caused by poor nutrition after emergence in the hive. Symptoms include larvae dying before capping in a twisted 'C' or halfmoon shape in the cell. Larvae change from white to yellow to light and then dark brown with tracheal lines still evident. The symptoms are similar to European foulbrood where larvae dry to form a 'scale' that is easily removed unlike American foulbrood scale that is difficult to remove. Many cells contain multiple eggs with eggs attached in chains and eggs laid on the wall of the cell. Often drone eggs are laid in worker cells. The queen may be superseded given she is failing.

Varroa mite (*Varroa destructor*): The reproductive rate of varroa on drone brood is 2.6 times compared to 1.3 times on worker brood. Drone brood is therefore 8 to 10 times more likely to be parasitised than worker brood. Only 60% of drones from varroa infested colonies were alive one day after emergence compared with 97% for non-infested colonies. Drones emerging from infested pupae often appeared normal but the majority could not fly and only 20% of drones reached sexual maturity (12 days old) compared to 37% from non-infested colonies. The amount of sperm produced by drones emerging from pupae infested with more than 3 mites was 4.3 million compared with 8.8 million sperm produced from drones emerging from non-infested pupae. Varroa mites feeding on worker pupae may result in smaller hypopharyngeal glands developing and hence a reduction in royal jelly production in nurse bees.

Parasitic Mite Syndrome (PMS) is the term used for abnormal brood and adult symptoms found in varroa infested colonies. Such symptoms include reduced hive population, crawling bees at the hive entrance, queen supersedure, sunken perforated cappings, twisted 'C' shaped larvae, larvae filled with a watery fluid lying along the lower wall of the cell that turn light brown to black eventually drying to a 'scale' that is easily removed.

CHAPTER TWO: QUEEN BEE REARING

2.1 Queen Rearing Equipment

2.1.0 INTRODUCTION

A colony in the wild can produce a new queen without human intervention so long as a fertilised egg is present. Beekeepers have developed techniques to rear large numbers of queen bees so that beehives can be requeened regularly (one to two years) to reduce swarming, increase honey production, start new colonies, or add other genetic improvements to a colony such as disease resistance.

The key to queen rearing is to take a female larva, from a worker cell, at a very young age (12-24 hours old) and place (graft) the larva into a queen cell cup suspended vertically in a strong hive. The larva is fed on royal jelly by worker nurse bees and in ten days time is ready to be placed into a hive.

A variety of specialised equipment is used during the queen bee rearing process. Most of the equipment is relatively low cost and can be re-used or constructed by the beekeeper. One of the most costly items in queen rearing is the nucleus hive that may vary enormously in size.

2.1.1 CELL BARS AND QUEEN CELL CUPS

Queen cell cups are plastic or wax cups about 8mm in diameter at the rim. Cell cups are attached to wooden or metal bars (cell bars) that are slotted into a frame of standard dimensions (figure 2.1.1(a)). Cell cups (20-60 per frame) are suspended vertically from the cell bars.

Female larvae are transferred using a grafting tool into the cell cups then suspended upside down in a strong hive. Cell cups are positioned in the brood area of a strong hive in a vertical position.

Most queen breeders use a standard frame that holds two or three cell bars. The bottom bars of standard frames can be made into cell bars by cutting them slightly shorter to fit inside a standard frame.

When using short Bozi cell cups (26mm) it is possible to have three cell bars in a frame, each one being approximately 55mm apart. When using longer (29- 30mm) Bozi cell cups, only two cell bars are used otherwise there is insufficient room between the cell bars for fully mature queen cells to develop.

Small strips of wood are nailed to the inside of the side (end) bars (figure 2.1.1(a)) to create a groove for the cell bar to fit into. When wax is coated on the cell bars, they will fit into these slots without falling out. Alternatively the cell bars can be attached by one nail at each end so that they swivel for grafting.

Plastic cell cups can be purchased from most beekeeping supply outlets usually in lots of 100. Previously used cell cups should be cleaned by scraping out royal jelly from the base of the cups then washing the cups in warm water with a little detergent. The cups should be left to dry out thoroughly before attaching them to a cell bar. Alternatively attach the cups to the cell bar, pour hot water into the cups, shake the water out, let them dry and then introduce into a hive.

Plastic cups are attached by melting clean wax in a pot or fry pan. Using a wooden or metal spoon, dip the spoon in the melted wax and pour the wax along the length of one side of the cell bar. Pour several layers of wax on the cell bar. Dip the base (widest end) of each plastic cup into the melted wax then push the cup into the layer of wax on the cell bar. Attach 15-20 cups along the cell bar (figure 2.1.1(a)) then pour a little extra hot wax around the base of the cups to ensure they are firmly attached. Dip the rim of the outside four cell cups at each end of the cell bar into wax to increase the acceptance of grafted larvae.

Cell cups can also be constructed of wax. Wooden dipping sticks are used as a mold. They can be constructed from wooden dowel by filing down the tip so that it makes a snug fit into a plastic cell cup (figure 2.1.1(b)).

Any number of tapered dowels can be joined to a wooden bar so that the cups are 18-20mm apart, centre to centre. The tip of the dowels are dipped in warm soapy water with detergent then dipped in liquid wax to a depth of about 10mm. The dowel is quickly removed; the wax allowed to cool, then it is re-dipped to a slightly lesser depth

Figure 2.1.1(a) Cell bars and frame with long Bozi cell cups on the top cell bar and short Bozi cell cups on the middle cell bar. Small strips of wood are nailed to the inside of the side bars to create a grove for the cell bars to fit into.

Figure 2.1.1(b) Wooden dowels used for the construction of wax cell cups.

and removed. This process is repeated three or four times to build a layer of wax around the dowel.

Finally the tips of the dowels are pressed on to a cell bar coated with a layer of molten wax. The dowels are removed from the wax cup by submerging the tips of the dowel, the cell bar and the wax cups in cold water. Individual wax cups can be made by dipping a single dowel in warm soapy water with detergent, then into the hot wax, then twisting the cup off by hand while the wax is still warm. If the wax cups do not twist off easily try smearing detergent around the dowel tip before placing in the hot wax.

2.1.2 GRAFTING TOOLS

A variety of different types of grafting tools have been used for removing larvae from worker cells and placing into queen cell cups. These range from tapered matchsticks, toothpicks, fine tipped paint brushes to fine metal or plastic spatulas and Chinese grafting tools (figure 2.1.2).

The end of the tool, the tongue or spoon, should be moist and smooth, flat and tapered to slide under the larva without rolling it over. A fine paintbrush either 00 or 000 is suitable; alternatively taper the end of a matchstick with a pocketknife.

The Chinese grafting tool is possibly the preferred instrument for grafting. This has a flexible plastic 'tongue' with a wooden shaft on a spring-loaded rubber end piece. When the rubber end is depressed the shaft pushes a wooden tapered end piece along the tongue to dislodge the larva. The advantage of the Chinese grafting tool is that a larva, once removed from a worker cell, can be easily dislodged from the tongue end into a cell cup by pushing down on the spring that extends the wooden shaft. This makes placement of a larva with royal jelly into a cell cup, much easier.

It is important to keep the tongue covered with a plastic protection cover when not in use, to prevent damage. Chinese grafting tools are available from some beekeeping equipment suppliers. All grafting equipment should be regularly sterilised in alcohol to avoid spreading diseases. If the plastic tongue becomes bent it can be straightened by placing in hot water.

2.1.3 CLOAKE BOARD, PHEROMONE EXCLUDER AND QUEEN EXCLUDER

Rearing new queen 'cells' requires isolation of the queen in an area of the hive so that access to the cells is prevented. After the larvae have been grafted into queen cups, the frame of cell bars can either be placed into a queenless 'starter hive' or

Figure 2.1.2 Grafting tools: top - OO paint brush; middle – Chinese grafting tool; bottom – tapered matchstick.

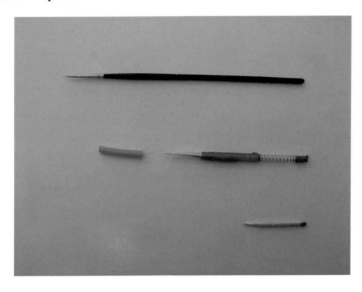

Figure 2.1.3 Cloake board (top left), queen excluder (top right) and pheromone board (bottom).

alternatively the cells can be placed above a queen excluder in a double storey 'cell builder' hive.

The 'queen excluder' has wires or slots to allow workers to pass through but the queen cannot, being wider in the thorax and abdomen. Normally the queen would be confined in the lower brood box with the queen cells placed in the upper box. Some beekeepers prefer the reverse with the cells in the bottom but this makes for more lifting of boxes to remove ripe queen cells.

The 'Cloake board', invented by Harry Cloake from Timaru, New Zealand, consists of a wooden board or metal plate that slides on wooden grooves. Cloake boards can be constructed using hive mats (inner covers) by removing one end of the inner cover frame and the plywood sheet. The remaining three sides of the inner cover frame are nailed to the top of a queen excluder. A metal sheet (same dimensions as the plywood inner cover) with one end bent over at right angles for approximately 10mm is constructed so that it slides into the grooves created by the inner cover frame (figure 2.1.3). Cloake boards are placed between the two brood chambers above the queen excluder.

The purpose of the Cloake board is to temporarily divide the two brood boxes in half so that no contact can be made between bees in the top and bottom boxes and no pheromones can be transferred from the queen in the bottom box into the upper box where the queen cells are located. By sliding the Cloake board into position, bees in the top box effectively become queenless as they cannot make contact with the queen. By removing the Cloake board with the queen excluder still in position, worker bees develop a supersedure response making contact with the queen. However the queen cannot make contact with the queen cells.

The 'pheromone excluder' is made up of a queen excluder with a rectangular piece of hardboard (390 x 310 mm) fixed to the excluder at each corner. Worker bees can still move through the outside of the excluder but the movement of pheromones from the queen (queen substance) is restricted by the hardboard sheet.

2.1.4 QUEEN CELL INCUBATOR

Modified egg cabinet incubators used to hatch chickens are suitable for incubating queen cells during the pupal stage. Queen cells are removed from the hive after capping. A bowl of water is placed in the bottom of the incubator to maintain the humidity above 50%. The cells can be left on the cell bars and suspended in the incubator. The temperature is maintained at 34°C and 60-70% relative humidity and

Figure 2.1.4 Portable queen cell incubator that can be plugged into a cigarette lighter socket in a vehicle.

Figure 2.1.5 Emergence cages for individually containing emerging queen bees.

the cells are removed ten days after grafting. Taking capped queen cells out of the hive and keeping them in an incubator frees up the hive for rearing further larvae through to the capped over queen cell stage.

Portable queen cell incubators (transporters) are available from beekeeping supply outlets and will carry either 70 or 144 queen cells. The portable incubators are modified from small metal toolboxes and have a hot plate that is heated via a power source connected to the cigarette lighter socket inside a vehicle. A small light indicates the incubator is heating. A second light remains on until the incubator reaches the required 34°C. A 50mm thick piece of foam sits on top of the plate and has a series of 15mm diameter holes for placing the queen cells into (figure 2.1.4). When the cells are in place, the lid is closed to retain the heat.

The incubator should be connected to the cigarette lighter for about 15 minutes prior to introducing the queen cells, and a colour change on the temperature scale on the hot plate will show when the incubator has reached 34°C. The incubator should not be left in direct sunlight for too long as overheating may kill the queen cells.

The alternative to using a portable incubator for transporting cells is to place the frame of mature queen cells into a nucleus hive with plenty of bees. Place the mature queen cells between two brood frames to keep the cells as warm as possible. Queen cells can be individually removed from the frame and placed into queenless hives as required.

2.1.5 EMERGENCE CAGES AND CELL PROTECTORS

Emergence cages are used for emerging queens. If a queen cell appears to be hatching prematurely the cell can be placed into an emergence cage where the queen can hatch and be contained without destroying other queen cells (figure 2.1.5).

Cell protectors can be placed over queen cells either to prevent them emerging or to allow the queen to emerge without the cell being chewed down around the sides by another queen and workers in a queenright hive. Insulation tape, tin foil and plastic tubing can be used as cell protectors.

2.1.6 PROTECTIVE EQUIPMENT

Good protective clothing is essential for queen bee rearing as hives are being constantly manipulated. Some protective gear, such as gloves, can be left out as more experience is gained. Protective equipment is available from bee equipment stockists. Good equipment may cost more to purchase, but it is important to remember that if used and stored correctly the equipment will last for a long time. The protective gear required by a queen bee rearer includes the following:

BEE SUIT: A full pollination suit provides the best protection; this includes overalls and veil in one. The veil should sit well away from the face (figure 2.1.6). The suit is worn over a layer of clothing, including long trousers and long sleeved shirt for added protection. Alternatively, overalls can be worn with a half suit worn over the overalls. Bee suits should preferably be white as this is less attractive to bees and is also cool. The veil should be kept well away from hot objects such as the smoker as this may puncture the mesh.

BOOTS: Gumboots or steel capped boots are the best. They can be made bee-tight by tucking the trouser leg into socks with the overall leg on the outside. The bottom of the overalls should have elastic to hold them firmly against the boots, or tramping gaiters to keep them tightly secured to the boots. Steel cap boots are preferred if lifting honey supers.

GLOVES: Rubber gloves with long gauntlets provide good protection but are difficult to manipulate, whereas leather gloves have good manipulation for queen bee rearing but do not offer the protection provided by rubber gloves.

HIVE TOOL: A hive tool is a toughened steel lever used to prise apart hive boxes, or for separating one frame from another. Hive components are often stuck together with wax or propolis and the hive tool, with its flat blade and sharpened end, allows hive parts to be separated.

The hive tool comes in many different designs, with the most effective having a right-angled hook at the bottom for lifting out frames. The hive tool should be carried in the narrow side pocket of the overalls. If the hive tool comes into contact with American foulbrood it can be sterilised by scraping the wax and propolis off and dipping in a 0.5% sodium hypochlorite-based product e.g. Janola, for 20 minutes. An alternative method is to scorch the hive tool in a hot flame. Wash the honey and wax off the hive tool, in a bucket of water, between hives.

Figure 2.1.6 Protective equipment: pollination suit, gloves, steel cap boots, hive tool and smoker.

SMOKER: Smoke from the smoker helps to calm the bees and the bees react by filling their honey stomachs with honey from open cells, in preparation to leave the hive. With a full honey stomach workers are less able to sting. In fact it has been shown that a queen in full lay would not be capable of rapidly leaving a hive, as she would need several days to taper off egg laying to the stage where her abdomen would be small enough to allow her to fly.

Smoke puffed into the entrance of the hive will disorientate the guard bees. The simple method of smoking a hive a few minutes before opening gives the maximum effect. If you are checking a number of hives or nucs it is a good habit to lightly smoke several hives ahead of the one you are actually working on.

The smoker is made up of three major parts:
- Snout and lid
- Fire box
- Bellows

The fire is lit inside the fire box, on top of a small grate. A 97mm diameter fire box is preferred as this allows the smoker to continue burning for long periods with the correct fuel. Fast burning fuel, such as dry leaves and hay, wrapped in slow burning fuel such as hessian sacking - to create a fuel capsule - will produce the desired effect. Compression of the bellows pumps air into the fire box and forces smoke out of the snout. The smoker when lit should produce a cool, dense smoke, rather than flames which will burn the bees' wings and make them aggressive. When not in use, block the snout with newspaper from the inside ensuring that no air can get through. Or use a cork or plastic queen cell cup and make sure the lid is firmly on. Store the smoker in a fireproof metal box with a metal lid.

2.1.7 HIVE COMPONENTS FOR QUEEN REARING

1. Base

The base is the foundation or support for the hive. Onto the base the brood boxes are placed. The base has to be solid so as to adequately support the hive. The base consists of two runners (40x50x405mm) that sit on the ground and are flush with the front and rear of the base. Attached to the runners is the bottom board, which is the same width (405mm) as the brood box but is 30mm longer (535mm). Bottom boards, without runners, last no more than one or two seasons. The bottom board has 10mm high and 20mm wide wooden risers on the sides creating an entrance at the front and rear once the first brood box is placed on top. The base will have one entrance open and the other closed with newspaper, or with an entrance closer. During queen rearing the front entrance is blocked and the back entrance is opened so that the entrance faces the opposite direction to its normal position.

2. Bottom brood box (full depth)

The bottom brood box (405 x 505mm) usually contains nine or ten frames and includes the queen and the remaining frames not required for the top brood box. The queen bee lays eggs in these frames which develop through various life stages emerging as adults. Some of these frames may be transferred to the top brood box if required. The queen is confined to the bottom brood box by the queen excluder.

3. Queen excluder

The queen excluder is made of wire or plastic mesh sometimes with an outer wooden frame. The spaces between the mesh allow worker bees to pass through but not the queen bee or drones. The queen excluder is placed between the bottom and top brood boxes.

4. Cloake board

The Cloake board has a wooden frame similar to a hive mat and is nailed to the top of a queen excluder. One end of the frame is removed and a wooden or metal sliding panel (the Cloake board) is inserted with either a folded lip or an extended lip at the open end. The Cloake board is inserted into the grooves in the wooden frame so that it effectively seals off the top brood box from the bottom brood box. The Cloake board and queen excluder are placed above the bottom brood box so that when the top brood box is introduced there is an upper entrance at the front of the hive (figure 2.1.7(a)).

5. Sloping board

A wooden or metal board is placed in the front of the hive to provide a platform for bees returning to the front of the hive looking for a bottom entrance. The board provides a landing platform where bees can make their way to the front top entrance.

6. Top brood box

The full depth top brood box sits on top of the Cloake board and has an entrance to the front created by the Cloake board. This box is queenless and contains the queen cell cups in the middle frame. On either side of the cell cups are two brood frames with 3-5 day old larvae. Outside these two frames are two pollen frames. Outside these two frames are two unsealed honey frames and then, of the two outside frames, one may contain empty drawn comb and the other unsealed honey.

Figure 2.1.7(a) Hive set up with Cloake board in place and sloping board up to the new top entrance. The original entrance is at the back of the hive.

Figure 2.1.7(b) Demaree board opening (with swivel entrance) and division or split board (with 30 x 10mm entrance).

7. Honey super

An optional three-quarter or full depth honey super may be placed above the top brood box. This provides additional storage area for sugar syrup but has the disadvantage of having to be removed in order to remove the queen cells. Additional honey supers are only required if rearing queen bees during summer.

8. Feeders

Top feeders are placed above the brood box or honey supers. The feeder has a wooden outer frame (505 x 405 external dimensions) with a plastic or tin inner lining and has branches, dry bracken fern, corks or other floating devices placed inside to prevent bees drowning as they feed on the syrup. Sugar syrup (50:50 sugar: water) is poured into the feeders to stimulate the queen to lay eggs, or workers to feed royal jelly - so the feeders must not leak. The feeders can be easily filled without disturbing the hive simply by removing the lid and inner cover and pouring the syrup into the basin. Bees crawl up through holes in the middle or up the outside of the plastic inner container to access the sugar syrup without exiting the hive.

Frame feeders are commonly made of plastic and fit into the brood box. They replace 1, 2 or 3 frames depending on their width and may be used as an alternative to top feeders. They are usually filled with bracken fern.

Another feeder is the **inverted container feeder**. This consists of an empty super and a 100-200mm diameter tin and lid, with ten, 1-2mm holes punched in the lid. The tin is filled with syrup and the lid firmly replaced, then turned upside down and placed on top of a hive mat with a 100-200mm diameter hole. The tin is surrounded by an empty super and placed above the brood area. The major drawback is that this method requires a hive mat with a large hole and an empty super.

The **Boardman feeder** is an upturned glass jar filled with sugar syrup with 1mm holes in the lid, placed inside a wooden base, with an entry point for bees on one side only. The entry is slid into the hive entrance so bees access the sugar syrup in the jar at the entrance to the hive without exiting the hive, thus preventing bees from other hives gaining access. The amount of syrup remaining can be easily determined, without opening the hive, from the height of syrup in the jar. These feeders are often used for nucleus hives especially in caged pollination trials.

9. Inner cover or hive mat

This is a wooden hardboard sheet with a wooden outer frame (505 x 405mm) placed under the lid to prevent it being 'glued' down with propolis.

10. Division board and Demaree board

A division (split) board is similar to a hive mat except that at one end or side there is a 10 x 30mm entry/exit for bees cut out of the frame (figure 2.1.7(b)). The division board is used to divide a hive into two by splitting the brood. It acts as an inner cover for the brood box below and an entrance for the box above.

The Demaree board can be made of plywood or galvanized iron and looks similar to the division board except that it has an 80mm wide swivel (cut at 45° angle) entrance (figure 2.1.7(b)) that can be opened to provide an exit point for a division, or if closed, to lock bees in temporarily after a division has first been made up.

11. Lid or cover

The lid is usually telescopic i.e. fits over the brood box on all sides. It has a metal cover and wooden inner. The lid protects the hive from rain and from robber bees.

2.1.8 NUCLEUS HIVE

The nucleus hive (plural is nuclei; or abbreviated to nuc) is normally the same height and length as the standard full depth hive, but it is only half the width, or less. The most common nucleus hive in New Zealand is a full depth box that takes five frames (figure 2.1.8). In Australia three to four frame nucleus hives are more common. The base is usually attached to the hive box and the nuc has a removable telescopic (overlapping sides) or migratory lid (the same width as the box).

The nucleus hive may have a telescopic lid and a hive mat that can be made of plywood or linoleum (lino). The hive box or lid would normally have a ventilation hole covered in fly wire.

The entrance is small (10 x 30 mm) and may have the addition of an entrance closure or metal disc that provides either an open entrance, a closed entrance, a queen excluder entrance, or an air vent option to be 'dialed'. The entrance may be decorated with different colours and/or symbols to assist the queen bee in orientating to her hive during mating flights.

Multiple nucleus hives are constructed from a normal ten-frame full depth hive divided into three equal compartments using partition boards. The base has risers that have three entrances, one in front and one on each side, and effectively three nucleus hives each with three frames contained within one full depth box. Two partition boards slide

Figure 2.1.8 Nucleus hive (left) with telescopic lid, four frames and a single frame feeder. Polynucs (right) with three frames and a feeder, the lids are leaning behind the nuc and each has a plastic cover leaning against the side, the polynucs can be filled with bees from underneath.

Figure 2.1.9 Mailing cages: left to right, Miller cage; wooden mailing cage; plastic mailing cage; top opening plastic mailing cage showing candy and bee compartments; queen's guard mailing cage with top pushed forward into queen catching position and two compartments at the top for candy and moistened sponge; JZ's BZ's cage for mailing or introducing inseminated queens.

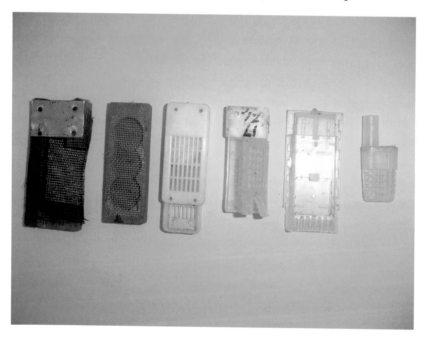

down a groove in the hive box and make contact with two risers, each with grooves that run the length of the base.

Some queen breeders will divide a standard full depth box into four compartments with two frames per compartment. Multiple nuc hives provide mutual warmth to each nuc compartment and are cost effective because they use standard size equipment. At the end of the rearing season the partitions can be removed allowing the hive to return to the standard nine or ten frame honey production unit.

Baby nucs, mini nucs or poly nucs (figure 2.1.8) are half the length of conventional nucleus hives with very small frames (approximately 125mm long x 24mm wide x 85mm deep) that run across the nuc box. These nucs are for mating only and survival of bees and the queen bee, for an extended period, is tenuous. Four mini nucs can be constructed from a standard size box (405 x 505mm) by dividing the box with two partitions - one running from end to end and the other from side to side. Four separate entrances are constructed from the risers on the base board.

Poly nucs have three frames and a feeder. The bees can be shaken into the nucs from the underside. The queen cell can be inserted in the dark and the feeder filled. A small queen excluder prevents the queen from entering the feeder. Air circulation is provided by an upper vent. The bees can be kept in the dark until the queen cell has emerged, then placed in the field and the entrance opened.

2.1.9 MAILING CAGES
A variety of different mailing cage designs are in circulation. These are used for transferring queen bees over short or long distances. Mailing cages provide a chamber for the queen and attendant escort bees, and a chamber of queen candy to supply food during transport.

Plastic mailing cages are around 23 mm wide by 70 mm long. The cages have two compartments (figure 2.1.9). A smaller compartment, 18-28 mm long and 23 mm wide, is used for filling with queen candy. There is a hole at each end of the candy compartment.

The external hole has a plastic seal that must be removed by the beekeeper with a sharp knife prior to introducing the cage into the hive. Removing the seal allows the queen to be released after a hole has been chewed through the candy by the workers.

The queen is placed in the mailing cage first, followed by 6-12 escort worker bees removed from the brood area. The cages have a sliding or clip on cover with slits for air circulation. Workers from outside the cage can pick up the smell of the new queen and if necessary feed her through the slits. The cages have four, 2mm plastic legs so that they can be stacked on top of each other with a 1 mm gap between for air to circulate.

Wooden mailing cages (external dimensions 35 x 90 x 19 mm deep) are made using three drill holes (of 26 mm diameter bored to a depth of 15 mm) with 8mm diameter holes in each end. One hole is blocked by queen candy. A cork is inserted into the candy end and sliced off flush, and the other end has a cork inserted once the queen and escorts have been introduced.

The top is covered with wire gauze and stapled to the wooden cage. The queen is captured with the thumb and forefinger and carefully encouraged through the hole at one end. Six to twelve escort workers are plucked off the brood by the wings and pushed through the hole. A small cork is then inserted and pushed into the hole as far as possible then sliced off flush with a razor blade.

Miller cages are made from wire gauze, bent over a wooden block of wood and are approximately 100 x 50mm. The cages are used for transporting queen bees' short distances within a beekeeper's operation. The queen and six to twelve escort workers are placed inside and the gap filled with queen candy. If the queen is to be transported only a short distance escorts are not required.

Queen's guard mailing cages are Israeli-designed rectangular cages (35 x 70 mm) with a sliding upper plastic cover. The top cover is clear and slides forward and backwards.

With the top cover in the forward position it may be placed over the queen and workers then slid back, capturing the bees inside without the need for handling by the beekeeper. There are two additional compartments. One is for queen candy and the other for a moistened sponge to supply water to the bees during transit.

The cages can be stacked with the help of a tiny hole on the bottom and a knob on top of each cage.

When the beekeeper receives the caged queen he can slide the cover forward to the 'bee only' position whereby escort bees are released without the queen exiting. The cover is then slid back into the 'release' position and the cage placed into the hive to be re-queened. Worker bees can chew through the candy and release the new queen. The cage has a spike to push into the brood and a hole where a nail can be pushed through to suspend the cage between two frames.

JZ's BZ's cages are used for mailing queen bees or introducing into nucs during the insemination process. The cages are filled with candy in the tubular end and slip easily between the frames. A small cap is removed in the main body of the cage to introduce the queen and then reinserted. There are vents along the sides to allow workers to feed the queen but which are small enough to prevent workers from biting the queen's feet.

2.2 Queen Rearing by Grafting

2.2.0 INTRODUCTION

Grafting is the transfer of larvae from brood frames to cell cups using a grafting tool. Producing good quality queen bees by grafting requires larvae from a breeder hive, that are less than 24 hours old. Grafting is a technique that needs to be practiced to acquire the necessary skills and speed. Well-populated hives are required to feed the grafted larvae to produce large queen cells and hence large, long-lived queens. The hives must be manipulated to first accept and then feed these queen cells.

2.2.1 BREEDER EGG LAYING

A queen bee breeder will select 5-10 'queen mother' hives with desirable characteristics from which to breed. To prolong her life and reduce egg laying demand, each selected 'breeder' queen should be placed in a 5-frame nuc or 10-frame single hive with plenty of pollen and honey.

Grafting several hundred larvae produced from a breeder queen, requires the correct hive manipulation for up to a week beforehand so the larvae are the correct age when grafted. The nuc should be supplied with 50:50 sugar: water using either a frame, Boardman or top feeder, two weeks beforehand. This feeding will stimulate the queen to lay eggs.

The queen bee requires a drawn, empty, dark comb placed in the brood area of the breeder hive in which to lay. Other remaining frames in the nuc should be full of brood, pollen or honey with few empty cells so the queen will concentrate her egg laying on the dark comb introduced. The comb should be warmed and polished by placing above the excluder of a well-populated hive with an active brood nest - for several days prior to inserting into the breeder nuc. This ensures the frame will be more acceptable to the confined queen for laying purposes. The hive should not be fed while the frame is being polished otherwise it will be filled with sugar syrup by the nurse bees.

2.2.2 CONFINING THE QUEEN BEE

The queen bee may be confined to the frame introduced, by partitioning off the area with two queen excluders, one on either side of the empty comb introduced, and an excluder above and below the frame. To do this make two saw cuts in the hive body about 50mm apart so that the queen excluders can slide down either side of the frame and make contact with the bottom board.

A bee space (6-9mm) should be left between the frame and the queen excluder on each side to ensure the queen is not crushed when the frame is removed. The breeder queen is confined for 24 hours. The frame can then be removed and placed outside the queen excluder with another empty frame introduced. The original frame should be checked for eggs and the date recorded.

Assuming the queen laid in the frame during the first day, the eggs will take three days to hatch, and 12-24 hours after hatching the larvae should be ready to graft. So, four days after the frame was introduced, assuming the queen laid eggs immediately after introducing the empty frame (this is not always the case), the larvae should be grafted into queen cell cups. For example, introduce an empty frame into the breeder hive on Monday morning and graft on Friday afternoon. This allows about 6 hours for the queen to start laying.

When the frame is removed for grafting gently brush the bees off with a fine bee brush. Shaking may disrupt the larvae. If only one frame of larvae is required, after 24 hours the queen excluders may be removed and sugar syrup feeding continued. A frame of pollen should be placed on one side of the frame of eggs, with a frame of mature larvae, containing an abundance of young nurse bees, on the other side. This will ensure that when the larvae hatch they will be supplied with copious amounts of royal jelly. If more than one frame of eggs is required, the first marked frame in which eggs have been laid should be transferred into another hive supplied with bees and food stores as above.

Another method of confining the queen is to provide a lay cage (100mm x 100mm with 18mm high sides, see Jenter or Ezi Queen method). The lay cage has a plastic queen excluder cover to allow workers to move in and out of the cage. The plastic cover fits over the lay cage to confine the queen. The cage can be used to confine a breeder queen to an area of empty drawn brood comb for 24 hours. Workers continue to feed the queen during this confinement period.

2.2.3 QUEEN CELL CUP PRIMING

To increase the acceptance of grafted larvae it is important to 'prime' the queen cell cups once they have been attached to the cell bar. Priming occurs when nurse bees clean and polish the queen cells with wax and propolis and often a small amount of wax is placed in a ring around the rim of each plastic cup (figure 2.2.3). The nurse bees also warm the cups. In a hive that has been prepared for feeding grafted larvae, it is important to place the cell cups in the middle of the brood for one or two days prior

Figure 2.2.3 Two bars of primed queen cell cups.

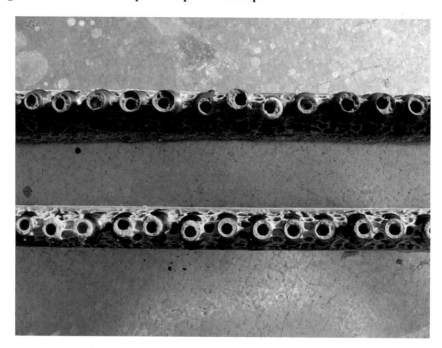

Figure 2.2.4 Inserting varroa strips into hives.

to grafting. Priming is essential if a high percentage of grafted larvae are to be accepted.

Dipping the last four cups at each end of the cell bar in molten wax - to provide a ring of wax around the cup - may also increase the acceptance rate.

2.2.4 VARROA MANAGEMENT

All colonies involved in queen cell production, and hives used for making up nucs, should be treated for varroa mites in spring (figure 2.2.4) and then again in autumn after nucs are united back to the parent hive. Treatments last for 6-8 weeks and should be removed before honey production. Treatments should be alternated to prevent resistance building up in the mite population. Sufficient drone mother hives (hives supplying drones for mating) need to be provided for mating as varroa mites can reduce the number of drones reaching sexual maturity. Drone mother hives should be treated 40 days prior to when the drones are needed for mating. Do not use formic acid as this causes drone eggs to be removed from the comb. Survey drone brood with a capping scratcher to determine what rate of infection is present. Survey 10% of support colonies using recognized sampling methods e.g. sugar shake, every two months. Carry out treatments whenever the population reaches 2,500 mites in the hive. Monitor the mite numbers after treatment.

2.2.5 FINDING THE QUEEN BEE

To find the queen in a strong hive to be used to rear queen cells, smoke the entrance several minutes before opening with a cool smoke to disorientate the guard bees and give sufficient time for other bees to engorge on honey.

Remove the lid, place upside down on the ground, place the inner cover to one side, remove any honey supers and place on the lid. Smoke through the queen excluder to drive the queen into the top brood box. The excluder is now removed and checked to ensure the queen bee is not present. The excluder is turned upside down and placed on top of the super(s).

The brood boxes are carefully separated and the top brood box is placed on top of the queen excluder. By separating the two brood boxes there is a greater chance of locating the queen, otherwise she can freely move between the two brood boxes. Scan your eye across the top of both brood boxes. Begin your search in the brood box where bees appear quietest; this is often an indication the queen is present. Cover the remaining brood box with the inner cover.

The procedure for finding the queen bee is as follows:
1. Using minimal smoke, slowly and carefully remove the second (from outside) frame from the quietest brood box, being sure not to crush any bees.
2. Hold the frame up on a slight angle.
3. Search around the outside of the frame for the queen bee.
4. Check along the bottom of the frame.
5. Check across the frame (figure 2.2.5), moving your eyes in a 'W' pattern.
6. Rotate the frame by bringing the top bar into a vertical position, and then rotate around its axis and bring the top bar down into a horizontal position.
7. If the queen bee is definitely not present on this frame:
   - lean the frame up against the hive with an end bar on the ground.
8. Repeat the procedure (1-6) for the remaining brood frames. The queen bee is less likely to be found on solid honey frames and most likely to be found on frames containing eggs or empty cells.

Place each frame that has been inspected back into the space provided by the previous frame removed. Finally, return all frames to their original position.

The second brood box (covered by the inner cover) is now checked, using the same procedure. If the queen bee is not located after searching the second brood box, check the sides of the box. If the queen bee is still not located, it may be necessary to perform the following procedure:
1. Place two empty hive boxes (with a queen excluder between them) on top of the baseboard.
2. Shake all the bees from the hive into the top box.
3. Drive the bees through the excluder, using a smoker, into the bottom box.

The queen bee should remain trapped on top of the excluder. This is a time-consuming process and probably should only be used as a last resort, as it causes considerable disruption to the hive and may initiate robbing.

2.2.6 REMOVING THE QUEEN BEE

To remove the queen without catching her, simply remove the frame that the queen is on and place this in an empty nuc. Remember to have a tight fitting lid or inner cover for the nuc, and block the entrance with newspaper.

To catch the queen without picking her up by hand, use one of the following methods:

1. A **queen catcher** looks like a plastic or metal bulldog clip with a chamber to hold the queen bee (figure 2.2.6). Capturing the queen using this method requires considerable practice to ensure that the queen bee is not injured.
2. A **queen holder** looks like a pair of plastic tweezers with sponge-covered tips that grab and hold the queen without injuring her.
3. A **queen's guard mailing cage** can be opened and placed over the queen while she moves on a frame; the opening is carefully closed catching the queen inside.

If the queen is to be picked up, gloves should be removed. If right handed, then it is easier to use this hand to pick the queen up.   Moving slowly with your hand behind the queen and with no jerking action, come down over the top of the queen with thumb and forefinger and grasp her firmly by the thorax; this will not squash her. Transfer the queen carefully into a mailing cage and slide the lid shut. The cage can be placed in a secure overall pocket until required.

2.2.7 QUEEN REARING IMPULSES

The emergency and supersedure impulses are the two queen rearing impulses most often used by the beekeeper in the production of queen cells in a queen rearing programme, although the swarming impulse can also be used.

Emergency response

Beekeepers use the emergency response for starting or initiating feeding of queen cells because a large number (20 – 60) of cells can be initiated using this response. A strong colony that is made queenless and deprived of young brood suitable for transforming into emergency queens, will readily feed young larvae in plastic queen cell cups provided by the beekeeper.

Removing the queen from a strong single storey hive, making it queenless, can create the emergency impulse. The alternative is to take a two-storey hive and place the queen bee in the lower box. The two boxes are then separated completely by sliding a panel (the Cloake board) between them. This effectively creates a queenless hive in the top box and therefore an emergency response.

The emergency response created in the top box by the absence of the queen means that the queenless workers will readily accept and feed queen cells. Workers in a 'queenright' hive (with a queen) will not always accept queen cells especially if the queen in the hive is less than a year old.

Figure 2.2.5 Locating the queen on a frame. The queen is marked with typist correction fluid so locating her is easier.

Figure 2.2.6 Queen catching devices: left to right, queen holder, queen catcher and queen's guard mailing cage.

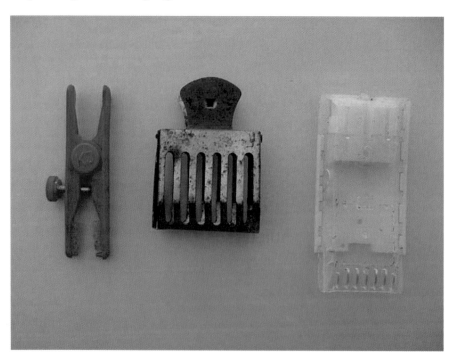

Although the workers will start feeding a large number of queen cells under the emergency impulse, most beekeepers prefer to have the queen cells finished under the supersedure impulse to ensure the cells have the best feeding possible for the longest time.

Supersedure response

To induce the supersedure response the queen should be in her second year. This is to ensure that there is not an abundance of queen substance passed through the queen excluder, and the workers in the upper brood box do not receive a full share of the queen's pheromones, hence they perceive her as failing. The reduction of queen substance in the top box induces a supersedure response in the workers.

Queen cells are placed in the upper brood box of the hive surrounded by worker brood (mature larvae), with the queen usually in the lower box beneath a queen excluder; this stops the queen reaching the new queen cells and destroying them.

Removal of the Cloake board 24-36 hours after introducing queen cells changes the queen rearing impulse. This removes the emergency response and induces the supersedure response with workers now able to make contact with the original queen. Under these conditions queen cells are fed more frequently and for longer before the cell is sealed so queen cells become larger and produce bigger queens.

Swarming response

The swarming impulse can be introduced into the queen rearing programme to enable sufficient nurse bees with well developed hypopharyngeal glands to feed royal jelly to the developing queen larvae. This will ensure there is never a shortage of royal jelly for developing larvae.

Crowding a three-storey hive into a two-storey hive, providing an over-abundance of bees covering the frames, can induce the swarming response.

2.2.8 ESTABLISHING CELL BUILDER HIVES

Many beekeepers will place grafted cell bars into queenless 'starter' colonies or a 'swarm box' for 24-36 hours in order that a high percentage of grafted larvae in queen cups will be fed and queen cell construction will be initiated. These queen cells, once accepted, are transferred into 'finisher' colonies where larval feeding continues until the queen cells are sealed over with wax (capped) and pupation begins.

As the queen spends only about 5½ days as a larva (4½-5 days in the unsealed stage), it is vitally important that the maximum amount of feeding is provided during this period. By the time larvae are grafted they may only have 3½-4 days of feeding before being capped.

Some beekeepers combine the starter and finisher hive into one 'cell builder' hive. This is possible with the use of a queen excluder and Cloake board. The cell builder hive is set up to feed queen cells copious quantities of royal jelly for approximately four days prior to capping.

The cell builder hive is usually a two-storey, full-depth frame hive. The hive(s) selected should be very well-populated with bees, to the point where they may be near to swarming (figure 2.2.8(a)).

Taking a three-storey hive and shaking the bees into a double hive for cell raising can achieve the swarming impulse. Nurse bees from the cell builder hive should have the capacity to feed copious amounts of royal jelly to the larvae. This can be determined by checking young larvae from a few frames to see if they have been adequately provisioned with royal jelly in the cell.

To set up a cell builder hive ready to accept queen cells, open the hive, and before shaking and disrupting the hive, locate the queen. A one or two year old queen is preferable to a new queen for rearing queen cells. The older queen produces less queen substance and so the impulse for rearing new queens in the hive is greater. The queen bee should be marked and temporarily placed in a queen mailing cage with a few workers and queen candy. All brood frames should be inspected to ensure there is no disease. Remove all the frames from the top brood box. The frame containing the cell bars is placed in the middle of the empty brood box. Two brood frames with no eggs or young larvae, and with mature worker larvae that fill the cells (3-5 days old), are placed on either side of the frame with the cell bars (figure 2.2.8(b)).

Two frames of pollen are placed outside the two brood frames. Two frames of unsealed honey containing fresh nectar or sugar syrup, are placed outside the pollen frames. One of the outer frames may be an empty drawn comb or replaced with a single frame feeder while the other frame can be unsealed honey or a frame that is partially empty, so that sugar syrup can be readily stored in the frame by worker bees.

Figure 2.2.8(a) Strong hive suitable as a cell builder.

Figure 2.2.8(b)

Set up of Cell Builder Hive using a Cloake Board

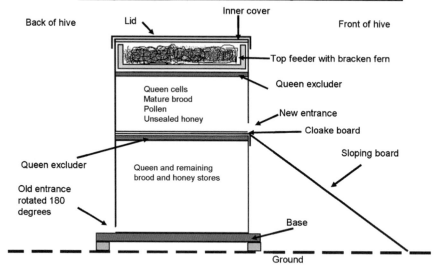

Back of hive — Lid — Inner cover — Front of hive

Top feeder with bracken fern

Queen cells
Mature brood
Pollen
Unsealed honey

Queen excluder

New entrance

Cloake board

Queen excluder

Queen and remaining
brood and honey stores

Sloping board

Old entrance
rotated 180
degrees

Base

Ground

Top View

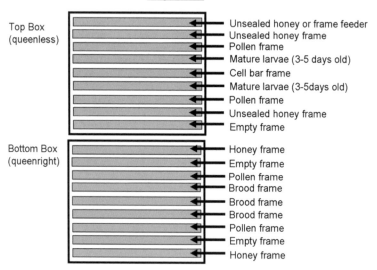

Top Box
(queenless)

- Unsealed honey or frame feeder
- Unsealed honey frame
- Pollen frame
- Mature larvae (3-5 days old)
- Cell bar frame
- Mature larvae (3-5days old)
- Pollen frame
- Unsealed honey frame
- Empty frame

Bottom Box
(queenright)

- Honey frame
- Empty frame
- Pollen frame
- Brood frame
- Brood frame
- Brood frame
- Pollen frame
- Empty frame
- Honey frame

As many bees as possible should be shaken into the upper box from the remaining frames. Place all remaining frames into the bottom box. Carefully reintroduce the queen into the bottom box using smoke to drive her down. Rotate the entire bottom box 180 degrees so that the original entrance is now at the rear of the hive. Place a queen excluder and Cloake board on top of the bottom box so that there is a new entrance created at the front of the hive between the first and second box. Place the second box, containing the queen cells, above the Cloake board. The queen is now separated from the upper box by a queen excluder and Cloake board, the latter is inserted in position to divide the hive into two separate compartments. No queen substance can pass from the queen in the lower box to the worker bees in the upper box.

The upper box should be covered by a hive mat (inner cover) to keep the cells warm. If a top feeder is used instead of a frame feeder this will be located above the second brood box and covered with a hive mat and lid.

The feeders should be filled with bracken fern or similar, to prevent bees drowning, and constantly supplied with 50:50 sugar:water syrup for two weeks prior to grafting, and for four days after grafting when all the queen cells will be capped over.

The hive should be supplied with a pollen supplement or substitute when grafted larvae are introduced. If bee products are used in supplements ensure there is no possibility of these products containing American foulbrood.

On the morning of grafting prepare the cell builder hive as previously described. Remember to rotate the bottom brood box 180° so that the old entrance is now at the back of the hive. Insert the Cloake board in position so that there is an upper entrance at the front of the hive. Place a sloping board in front of the hive from the ground up to the new Cloake board entrance so that workers leaving from the lower back entrance will return to the front entrance and eventually walk up the sloping board into the upper entrance above the Cloake board.

After about five hours of foraging many worker bees should have left the bottom queenright brood box and returned to the upper queenless brood box. By mid-afternoon the upper brood box should be packed with bees and a queenless 'roar' should be heard when opening up the upper box to remove queen cups for grafting. There should be a large number of worker bees hanging from the cell cups when the cell bars are removed for grafting. These bees are shaken back into the hive.

The emergency response created by this queenless situation will encourage the nurse bees to rapidly feed young larvae grafted into queen cups, and to accept and feed many queen cells. This situation occurs naturally when a hive becomes queenless. New queen cells are produced from young worker larvae by nurse bees that float the worker larvae on a bed of royal jelly to the outside of the cell and tear down nearby cells to produce vertical queen cells.

Under the emergency response young queen larvae are not fed as well over four to five days of feeding as they would be under the supersedure response. Therefore, in order to artificially create the supersedure response in the cell builder hive, the Cloake board is removed 24-36 hours after grafting and the back entrance is blocked. All foraging bees now leave by the front upper entrance. This means that the queen is still confined to the lower brood box by the queen excluder. Nurse bees, detecting the presence of queen substance from the queen, will continue to feed the queen cells more frequently and for longer under the supersedure response created, than if the Cloake board was left in place and the emergency response allowed to continue.

The acceptance of grafted larvae is determined 24-36 hours after grafting when the Cloake board is removed, by examining the queen cups to see if construction of wax queen cells has started. The nurse bees will begin building a wax queen cell about 5mm long extending down from the plastic cup. The grafted larvae will be provisioned with copious royal jelly. At this time surplus cells that have been accepted can be transferred to other cell builder hives so that each hive has no more than 20 cells to feed. These additional cell builder hives should be set up in the same manner as the original hive with the supersedure response introduced by removing the Cloake board and with the queen in the bottom box.

The queen cells are usually maintained in the cell builder hive until 10 days after grafting when they are removed. An alternative is to remove the accepted queen cells 36-48 hours after grafting (at the third day of larval development) and place these cells individually into well-provisioned queenless hives. The nurse bees will provision the larvae and cap the cell.

2.2.8.1 Two queen cell builder hive
A variation of the Cloake board method used by some queen rearers is to use two queens in the cell builder hive. The cell builder hive is established with three, three-quarter depth brood boxes. One queen is confined to the bottom box and a second queen is confined to the third (top) box with a top entrance. In between each of the boxes is a pheromone excluder. The cells are placed into the second box. The cells

are placed in the centre of the second box and surrounded by two frames of capped brood, and three frames containing pollen and honey are added outside of the brood frames. Two double frame feeders are placed on the outside of these six frames and supplied with sugar syrup.

2.2.9 GRAFTING LARVAE

Grafting is the transfer of larvae from brood frames to cell cups using a grafting tool. Grafting is a technique that needs to be practiced to gain accuracy and speed. Larvae to be grafted should be 12-24 hours old (figure 2.2.9(a)) and larvae older than 36 hours after hatching are unsuitable. The ideal age is larvae less than 12 hours old. These larvae are transparent and almost straight. The larvae should be surrounded by worker jelly, a milky white secretion from adult worker bees.

When grafting with a Chinese grafting tool (figure 2.2.9(b)), slide the flexible tongue down the side of the cell wall then pull back slightly on the shaft to push the tongue under a single larva and bed of worker jelly. Remove the tongue in a reverse action to the way it was inserted taking care not to let the larva touch the side of the cell.

If the larva rolls during removal, it may suffocate as the spiracles (breathing holes) may become covered in worker jelly. The larva has three spiracles along each side of the thorax and seven along each side of the abdomen. If the larva is rolled during grafting it should be discarded.

Place the tongue of the grafting tool, containing the larva, into the base of a queen cell cup. Push the end of the grafting tool with the index finger - the wooden shaft will extend and dislodge the larva and a little worker jelly into the base of the cell cup. Some queen bee breeders place a drop of diluted worker jelly into the cup before grafting the larvae. This ensures that the larvae will not dry out due to a lack of worker jelly.

Repeat the grafting procedure carefully and quickly until all the cell cups have a larva inserted. It is important not to let the larvae dry out or become too hot or exposed to direct sunlight so cover the larvae with a damp cloth as they are grafted. A warm room heated to at least 22°C with a pan or jug of boiling water providing steam to give at least 50% relative humidity will provide the right environment. Commercial queen breeders usually have a grafting room with heating and lighting provided.

A lamp and magnifying glass, or glasses, may assist in viewing the larvae. Lighting from immediately above the table where grafting occurs, directed on to the table, will

Figure 2.2.9(a) Frame with small larvae suitable for grafting surrounded by worker jelly.

Figure 2.2.9(b) Grafting larvae with Chinese grafting tool using sloping board.

assist in selection of appropriate sized larvae. The light source should not be too strong otherwise it may overheat the grafted larvae. A fluorescent lamp or cold light source is suitable.

The frame selected for grafting should slope towards the grafter at a 9° angle from the horizontal. This can be achieved by placing another frame, or piece of 20mm thick wood, under the bottom bar of the frame. The top bar should be closest to the grafter with the bottom bar raised. This way the grafter can see directly into the base of the cells with the aid of overhead lighting. The cells on a frame slope up at a 9° angle from horizontal, sloping the frame with the top bar towards the grafter will compensate for this angle.

The most common fault of grafting is to graft larvae that are too large. These larvae are older than 36 hours.

Larvae selected should be:
- almost transparent
- nearly straight (slightly 'c' shaped)
- about 12 hours old
- the smaller the better.

Any larvae larger than approximately 2mm will be too old. Once the larvae on the cell bar have been grafted, the date and breeder queen of the larvae grafted can be written on the cell bar and the bar fitted to the frame and inserted back into the cell builder hive. This should be done quickly to prevent the larvae drying out.

After all the larvae required are grafted the remaining larvae can be washed out of the frame with water and the frame re-used in another breeder hive.

2.2.10 SWARM BOX AND STARTER HIVE

The swarm box is a queenless hive used for rearing queen cells using the emergency response. A 'starter hive' is a single box, nine to ten frame hive, without a queen but with ample brood, food stores and bees. The swarm box is equivalent to a 'starter' hive although, if the swarm box is very strong, it may be used as a 'finisher' hive as well. A finisher hive is equivalent to a two-storey cell builder hive.

The swarm box is the same length and width as a standard four or five frame nucleus hive but is 1½-2 times the depth. The swarm box has ventilation 'windows' running the length of both lower sides covered in fine wire mesh and has a rebate for frames to be suspended just below the rim (figure 2.2.10).

Figure 2.2.10 Swarm box with cell bars in the middle and two single frame feeders on the outside.

Worker bees are shaken into the swarm box from one or more well-populated hives, making sure that the queen is not included. Grafted larvae in cell cups are suspended from cell bars in a frame that is placed in the middle of the swarm box. The swarm box is supplied with 50:50 sugar:water, in two single frame feeders placed on either side of the cell bar nearest the outside of the nuc box. In some cases two frames of mature unsealed larvae may be placed on either side of the queen cells but often no additional frames are added.

Queen cells may be started in the swarm box and then transferred to cell builder hives after 24-36 hours, or if the swarm box is very strong, left in there until the queen cells are capped.

Swarm boxes need to be continually replenished with workers shaken in from strong hives as old workers die. This means there is a lot of work maintaining these hives.

Starter hives are used in the same way, with cells placed in the middle of the hive. Because there is no queen, brood needs to be constantly added from other hives to keep the starter hive strong.

2.2.11 DOUBLE GRAFTING

The double grafting technique has the advantage of providing young larvae with copious royal jelly from a very young age. The disadvantage is that this technique requires two grafts compared to the normal one graft.

Double grafting begins by grafting young larvae one to two days old into queen cups on cell bars. These grafted larvae are placed into the top box of a Cloake board hive as in normal grafting.

After two days, when all the larvae have been fed copious amounts of royal jelly, the cell bars are removed and the top of each queen cell is sliced off with a razor blade or sharp knife. The larvae are then exposed and are individually removed by tweezers and discarded.

The royal jelly is then removed either with a plastic spatula or sucked up by vacuum using a manual or mechanical suction pump.

A few millilitres of water are added to the royal jelly to dilute it. The cell bar is placed back into the hive to be cleaned and primed for one day by the workers. The cell bar is then removed and a drop of diluted (with water) royal jelly is placed into the base of each cell cup.

A frame of larvae less than 24 hours old is then removed from a breeder hive. Larvae are grafted onto the droplet of royal jelly in each queen cup and the cell bar is returned to the cell builder hive.

A further step introduced by some beekeepers is to remove these larvae two days after grafting and perform a third graft of larvae less than 24 hours old, from a breeder hive. In most cases however, this third graft is not required unless the larvae in the second graft were too large.

Feeding the hives on sugar syrup and pollen substitute is essential, and mature queen cells are removed from the cell builder hive ten days after grafting.

2.2.12 CLOAKE BOARD MANAGEMENT METHOD

14 days before grafting (e.g. Tuesday)
Feed cell builder hive and breeder hive with sugar syrup (1:1 sugar:water by volume). If pollen is in short supply add cakes of pollen substitute to the above hives.

7 days before grafting (e.g. Tuesday)
Select an empty frame of dark (not black) drawn worker comb and place in strong hive in second brood box. Place the queen in first brood box with queen excluder between the two boxes. Do not feed this hive.

4 days before grafting (e.g. Friday)
During the morning remove the empty worker frame from the strong hive and place into the middle of the selected breeder hive. Ensure there are no other empty frames in which the queen bee may lay. Continue to feed cell builder hive. Record the dates of each activity in a grafting record diary (table 2.2.12).

2 days before grafting (e.g. Sunday)
Make up three cell bars each with 20 cups per bar. Place bars into frame and introduce into the brood area of cell builder hive with a frame of unsealed larvae placed on either side of the queen cups.

1 day before grafting (e.g. Monday)
Remove excess honey supers to force all bees into a double hive. This hive should be very well populated with nurse bee and maintained at a near swarming strength.

Table 2.2.12 Grafting record. An example of four grafts undertaken one week apart and grafting every Tuesday (Note: 28 days in February).

Graft number	1	2	3	4
Person grafting	Cory	David	Phillip	Andrew
Date empty frame placed into breeder hive	10-2-06	17-2-06	24-2-06	3-3-06
Date cell bar placed into hive for priming	12-2-06	19-2-06	26-2-06	5-3-06
Breeder hive number	Z1	Z4	Z7	Z10
Cell raising hive number	Z26	Z27	Z28	Z30
Date of graft	**14-2-06**	**21-2-06**	**28-2-06**	**7-3-06**
Number of larvae grafted (cell cups)	40	40	40	40
Date Cloake board removed and cell bars are separated into second cell raiser hive	15-2-06	22-2-06	1-3-06	8-3-06
Second cell raiser hive number	Z21	Z22	Z24	Z25
Number of larvae accepted	37	31	32	36
Percentage of larvae accepted	93%	78%	80%	90%
Date feeding finishes and check for rogue queen cells	18-2-06	25-2-06	4-3-06	11-3-06
Date checked for crowning	23-2-06	2-3-06	9-3-06	16-3-06
Date queen cells removed from cell builder hive	24-2-06	3-3-06	10-3-06	17-3-06

Key:
For determining dates:
Note: The grafting date should be decided upon in advance and then other activities should occur before and after this set date.

Date empty frame placed into breeder hive
for queen to lay eggs into   = 4 days before grafting

Date cell bar placed into hive for priming   = 1-2 days before grafting

Date Cloake board removed and cell bars
are separated into second cell raiser hive   = 24-36 hours after grafting

Date feeding finishes and check for rogue
queen cells   = 4 days after grafting

Date to check for crowning   = 9 days after grafting

Date queen cells removed from
cell builder hive   = 10 days after grafting

The workers should have demonstrated the ability to feed copious worker jelly to larvae.

Grafting day (e.g. Tuesday)

NOTE: If the weather is likely to be wet on this day, the cell builder hive can be set up the previous afternoon.

Prepare enough cell builder hives to ensure there is one cell builder hive for every 20 queen cells accepted after grafting. During spring a hive near swarming can feed 30 to 40 queen cells but an average hive can feed around 20 cells during autumn.

During the morning set up the cell builder hive (see section 2.2.8 establishing cell builder hive) with the queen in the bottom brood box and a queen excluder between the two brood boxes. In the centre of the top brood box place the frame of 20-60 queen cups.

Rotate the entrance 180° so the bottom entrance is now at the back of the hive. Insert the Cloake board above the queen excluder. Place a sloping panel at the front of the hive so that bees returning to where the old entrance was will find their way into the new top entrance.

Leave the hive set up for at least five hours of good foraging weather so that bees leaving the back entrance will end up returning into the top brood box via the front entrance.

After five hours remove the frame from the breeder hive, that was introduced four days before grafting, and brush off the bees. If the queen laid eggs on the afternoon of the day the frame was introduced (Friday) these eggs should have hatched and the larvae should be about 24 hours old or less, and have been fed sufficient royal jelly for grafting.

Remove one cell bar containing 20 queen cups from the cell builder hive. Graft 20 larvae into these cups; cover with a cloth. Return the cell bar to the cell builder hive immediately after grafting. Repeat grafting for remaining cell bars. Do not expose the larvae to direct sunlight.

Write on top of the cell bar the breeder queen from which the larvae were grafted. Make a diary note of the date of grafting and breeder queen used. Count ten days after grafting and record the date the queen cells should be removed and transferred into hives for requeening.

Continue with sugar syrup and pollen substitute feeding of the cell builder hive. Feeding of the breeder hive can be discontinued unless further grafting is anticipated the following week.

1-2 days after grafting (e.g. Wednesday evening or Thursday morning)
Approximately 24-36 hours after grafting, block the rear entrance to the cell builder hive. Slide the Cloake board out. This is important! If the Cloake Board is left in for more than 36 hours small queen cells will be produced. Maintain the feeding regime.

Transfer surplus cell bars to other hives so that approximately 20 queen cells, that have been accepted, are placed in each cell builder hive. Cell bars should be placed in the middle of the cell bar frame for maximum warmth and feeding.

1-4 days after grafting (e.g. Wednesday to Saturday)
Ensure that cell builder hives do not run out of sugar syrup or pollen substitute. If the wax covering the queen cells is light in colour it has been freshly secreted and the queen cells will have been well fed.

4 days after grafting (e.g. Saturday)
Once the queen cells are capped over (figure 2.2.12(a)) they can be removed from the hive and maintained at 34°C in an incubator until ten days after grafting. This frees up the hive for use in further cell raising. Alternatively, transfer all queen cells into one hive for incubation and separate the cells from the queen in the hive by a queen excluder so the queen cannot access the queen cells and destroy them. The queen cells should not be jarred during pupal development as the wings may be deformed.

Check all brood frames in the top box of the hive(s) containing the cells to ensure there are no rogue queen cells that have been constructed (figure 2.2.12(b)). Destroy any rogue queen cells found on the brood.

Cell builder hives should be kept warm with a lino mat placed over the top bars of the upper brood box, or bring the inner cover down under the top feeder. Sugar feeding can be discontinued.

9 days after grafting (e.g. Thursday)

Check queen cells to determine whether any have 'crowned' – i.e.where the wax has been removed from the tip of the pupa by the workers (figure 2.2.12(c)). If any cells have been crowned, remove these cells individually and place into separate emergence cages (see figure 2.1.5). The cages should be located in a nucleus hive away from the remaining queen cells. Failure to remove crowned cells may result in the emergence of a virgin queen. This queen will proceed to chew a hole in the side of the remaining queen cells (figure 2.2.12(d)) and sting the developing queen pupae. The workers will then remove and dismember the queen pupae inside.

10 days after grafting (e.g. Friday)

Ten days after the larvae are grafted into cell cups the ripe queen cells should be removed from the hive (figure 2.2.12(e)) and placed in a portable incubator (see figure 2.1.4)).

On the morning of the tenth day after grafting, plug in the queen cell incubator and ensure it has reached 34°C. Remove the frame of cell bars from the cell builder hive and carefully brush off the bees. If the cells are joined together with wax use a razor blade or sharp knife to carefully cut the wax between the cells to separate them.

Check the queen cells for black queen cell virus (BQCV) prior to placement into new hives by holding the cells in front of a light source (called 'candling') to determine whether a black ring is present around the tip of the queen cell. Those queen cells infected should be discarded.

This problem can usually be overcome by requeening the cell builder hive and allowing the hive to recover for three to four weeks or discontinuing with the use of the cell builder hive causing the problem.

Transfer queen cells individually into the incubator. Cells that are not symmetrical - where the tip is sunken or the cell is very small or damaged - should be discarded. Plug the incubator back in to bring it back up to 34°C. The queen cells should be maintained at 34°C until they are placed into hives.

If the cells are to be used to requeen queenright hives, wrap each queen cell in insulation tape, tin foil or plastic tubing, without crushing the cell. Leave the tip exposed then place the cell back in the incubator (figure 2.2.12(f)). Transfer the cells in the incubator to the hives requiring requeening. Lift any honey supers and place each cell beneath the queen excluder, between two frames in the middle of the upper

Figure 2.2.12(a) A successful acceptance of grafted larvae to produce 20 queen cells out of 20 grafted, the cells are capped over. The cell bar is located in the middle of the frame for maximum warmth.

Figure 2.2.12(b) Rogue queen cell (centre right) with the start of two queen cells to the left of the capped over rogue queen cell.

Figure 2.2.12(c) Crowning of five queen cells, nurse bees remove wax from the tip of the cell in preparation for queen emergence.

Figure 2.2.12(d) The middle queen cell on the bar has been chewed down by a virgin queen emerging too early.

Figure 2.2.12(e) Ten-day old queen cells ready to go out into hives.

Figure 2.2.12(f) Wrapping cells in tinfoil leaving the tip exposed and placing in a portable incubator at 34 degrees before placing out into queenright hives.

brood box of the hive to be requeened. Queen cells to be placed into queenless nucleus hives do not require wrapping. Each cell should be pushed into the middle of a brood frame between two frames of brood.

15 days after grafting (e.g. Wednesday)
Check hives to determine queen emergence. A hole should be chewed out of the tip of the queen cell. If emergence did not occur, check the queen cell for black queen cell virus and requeen with another cell as soon as possible.

2.2.13 DETERMINING QUEEN BEE PERFORMANCE
To determine whether the queen needs to be replaced, check her behaviour and egg laying. The queen bee is usually found walking over the brood, checking for empty cells to lay eggs into. Her legs and antennae should be undamaged, and she should have a long, tapered symmetrical abdomen. As the queen bee ages, her wings may become frayed at the tips. This may be an indication that she is old and needs to be replaced. However, the main indicator of her performance is her egg laying ability. If there are plenty of eggs and young larvae present, this would indicate that the queen bee is performing well. The mature capped brood should be even with few holes missed.

The queen bee begins her egg laying in the middle of the brood and gradually moves outwards in ever increasing circles. Brood will therefore normally mature and hatch from the middle first.

If there is a scattered brood pattern with many holes missed this could be an indication of reduced egg viability or inbreeding. If drone brood appears in worker cells or multiple eggs appear in a cell, this could indicate a drone layer or Halfmoon Syndrome. If the eggs do not hatch they may be sterile. If the queen has an abnormality her ability to lay eggs may be reduced. In all these cases the queen should be removed and the hive requeened with a young vigorous queen.

2.3 *Queen Rearing without Grafting*

2.3.0 INTRODUCTION

There are many techniques for rearing queen bees. Some require grafting of larvae and others avoid the need for grafting. Different techniques will suit different situations and will depend on the number of queen bees required and the skill of the beekeeper. This section looks at queen rearing techniques that do not require grafting and hence do not rely on good eyesight and hand-eye co-ordination.

2.3.1 SIMPLE REARING

Many commercial beekeepers, if they discover a queenless hive and they do not have a replacement queen, will transfer a brood frame containing eggs from another hive to the queenless hive, after a thorough disease check of both hives. This technique is better than leaving the hive queenless as the worker bees will eventually rear an emergency queen. Introduction of a mated queen however, increases the survival chances of the hive significantly. One of the simplest ways of rearing queen bees is to make a hive (or nuc) that contains eggs and young larvae queenless by removing the original queen.

With no queen in the hive the emergency response is initiated and nurse bees will begin to produce queen cells once the fertilised eggs hatch into young worker larvae. The problem with this technique is that it may be two weeks before the queen has emerged and is ready to mate and a further two to four weeks before she is mated and starts to lay.

The hive will therefore have a break of four to six weeks in brood rearing during which time the population and food stores may have dropped below the threshold level for the hive to survive.

Some commercial beekeepers will make up queenless nucs with eggs and young larvae at one site then transfer these to an apiary site at least 5km away. In areas where there are good food supplies this technique may be acceptable but in areas of marginal food supply the success rate is likely to be low.

The other problem in establishing hives in this manner is the stress placed on young larvae. This stress, due to temperature fluctuations and insufficient attention in a weak nucleus, may result in disease outbreaks such as sacbrood and chalkbrood developing in the larvae.

2.3.2 MILLER METHOD OF QUEEN REARING

There are several ways of undertaking the Miller method. The first method uses a newly drawn honeycomb that is cut length-ways about half way up from the bottom bar, with the bottom half being discarded. The frame of drawn comb can be cut across in a zigzag shape to increase the surface area for cell raising (figure 2.3.2).

The second method uses pieces of foundation wax approximately 60mm wide that are melted into the groove in the top bar of an empty frame. These pieces extend half way down the frame and taper to a point. Each piece is separated by about 10mm from the next piece with about six pieces along the frame. This frame is placed between two capped brood frames. Other frames of pollen and unsealed honey are placed on either side of the brood frames into the second box of a hive that is fed with 50:50 sugar:water syrup.   The bees will draw out the foundation and the queen will eventually lay eggs in the empty cells.

Three days after the queen has laid eggs in the cells the Miller frame should be removed, the lower edge trimmed, and two out of every three eggs (with their cells), destroyed so that queen cells made from the remaining eggs will be separated from each other. The Miller frame should be transferred into a cell builder hive with a Cloake board in place. The cell builder hive should be prepared in the morning and set up as for the Cloake Board method. The Miller frame is transferred into the middle frame in the top box of the cell builder hive after the top box has been queenless for at least five hours. After one or two days, once the queen cells have been initiated, the Cloake board can be withdrawn to change from an emergency to a supersedure response. Ten days later these queen cells will be ready to transfer into queenless hives.

2.3.3 ALLEY METHOD OF QUEEN REARING

This method involves placing a frame of newly drawn comb into a breeder hive between two frames of sealed brood. The selected breeder queen should lay in this frame over the next 24 hours. Four days later the eggs will have hatched into young larvae.

The frame of young larvae is removed and, using a sharp knife dipped in hot wax, strips of brood, about three cells wide, are cut to a length of 200mm. Strips should be cut right through the wax. One row of cells is left intact then every second and third cell is destroyed leaving 15-20 cells, spaced 10mm apart, still intact. The strip of cells

Figure 2.3.2 Frame cut to use for Miller method of non-grafting. Note the difference in size of the drone cells (top left of frame) and worker cells.

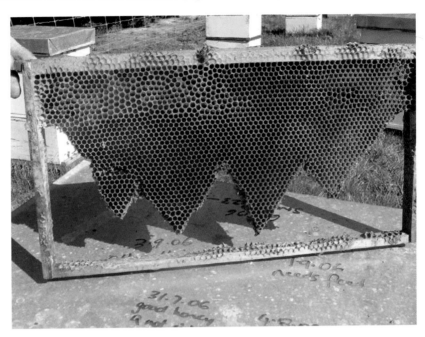

Figure 2.3.4 Ezi queen components: bottom-lay cage fixed to three quarter depth frame with queen excluder cover and plug inserted; middle – cell bars used for royal jelly (bottom bar) and queen cells (top bar); top – left 5 queen cups and two rows of ten plastic pegs that fit into back of lay cage, middle is the cover for back of lay cage, right is the back of lay cage showing rows of pegs.

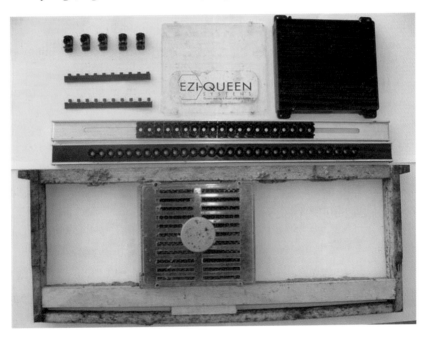

is attached to a cell bar using melted wax. The wax should not be too hot or this may damage the larvae in the cells.

The cell bar is fixed to an empty frame and placed into a cell builder hive in the afternoon after the Cloake board was introduced at least five hours previously. The Cloake board is removed 24 to 36 hours later.

Alley, in his 1885 book, suggested that a good cell building colony may start up to 25 queen cells but he advocated destroying some cells so that the colony would finish only 12 cells.

After ten days the queen cells are ready to be placed out into nucleus hives.

2.3.4 EZI QUEEN AND JENTER METHOD OF QUEEN REARING

These two similar systems are based upon the queen laying eggs onto plastic plugs that are removed from the back of a plastic lay cage (dimensions 132mm x 116mm for Ezi queen system). The plastic lay cage has four 'wings' - two on the top and two on the bottom - that slide into the underside groove in the top bar and grooved bottom bar of a three quarter depth frame (figure 2.3.4). Up to three lay cages can be secured into a three quarter depth frame. The lay cage contains 420 cells in which the queen may lay made up of 21 rows with 20 cells per row.

The lay cage is warmed and, with a soft brush, a thin layer of liquid beeswax is applied to the cells of the lay cage.

At the rear of the lay cage are 420 holes. Taking two strips of snap off plugs locate the lugs and dove-tail the strips together to form one strip of 20 plugs which are inserted into the first row of holes. Fill all 21 rows with plugs and secure the plastic back cover. The plugs form the base of the cell into which the queen will lay an egg.

A frame with one to three lay cages is then placed in a strong hive, in the middle of a honey super with capped honey, immediately above the brood. The hive is fed 50:50 sugar:water syrup and left for several days until the lay cage has been drawn out by nurse bees forming 420 uniform worker cells. If the lay cage is not drawn out after several days, attempt to confine more bees to fewer frames with the frame containing the lay cage being the only unsealed frame. Brush sugar syrup over the cells in the lay cage to attract nurse bees and continue feeding.

Once the lay cage is fully drawn, fit the plastic cover on the front, insert the breeder queen and secure her inside the lay cage by inserting the 30mm diameter front cover plug. Place the lay cage into the breeder hive. 'Escort worker bees' that move through the queen excluder slots in the plastic cover, will still attend to the queen.

During the next 24 hours the queen should lay 420 eggs in the drawn cells. The queen is then removed and can be placed into the next adjacent lay cage. Three days after the eggs are laid, the eggs will hatch into larvae and are ready to be transferred onto the cell bar.

Take an aluminium cell bar provided and bend the ends at right angles then fit the bar into an empty wooden frame by sliding nails into the holes provided. Slide or twist 20 plastic queen cups onto the cell bar and push together.

Remove the lay cage from the breeder hive and remove the back plastic cover. Remove the strips from one row, divide into two strips each with ten plugs, and insert one strip of ten plugs into ten cell cups. Snap off the plugs then repeat with the next strip of 10 plugs. Evenly space the 20 cell cups along the cell bar to prevent waxing and allow for queen cell protectors to be inserted. The transfer process is quick and up to 90 larvae per minute (420 total) can be transferred to cell bars. The Jenter system is similar to the Ezi queen system except that less larvae can be transferred from the lay cage to the cell bars at any one time.

Once the cell bar has 20 cell cups with newly emerged larvae it is placed into the top box of a cell builder hive. The cell bar is treated the same as a bar of grafted larvae, with the Cloake board inserted initially and then removed after 24-36 hours. If the transferred larvae have been accepted the nurse bees will begin building a wax queen cell about 5mm long extending down from the plastic cup. The hive is fed with pollen substitute and sugar syrup until the queen cells are capped (4-5 days).

After ten days the ripe queen cells should be removed by twisting the cells at right angles off the cell bar and placing into an incubator for distribution to queenless hives.

2.3.5 ROYAL JELLY PRODUCTION

Royal jelly can be produced using the grafting technique or without grafting using the Ezi queen or Jenter system. With royal jelly production, larvae are grafted 2-6 hours after they hatch and 72 hours (± 6 hours) later (three days) the royal jelly is harvested. The difference between producing queen cells and producing royal jelly is that once the larvae reach three days of age the cell bar is removed from the cell builder hive. The bees are brushed off the cell bar, tweezers are then used to open up the queen cell and remove and discard the larvae. The royal jelly is then removed either manually with a spatula or mechanically using a pooter (suction apparatus). The royal jelly is squeezed through a sieve (0.18mm diameter) at the base of the pooter into sterile glass vials that are then frozen to maintain the royal jelly properties.

2.4 Capture and Transport of Queen Bees

2.4.0 INTRODUCTION

Once the queen bee has successfully mated, the next important task is that of finding her, marking, possibly clipping the wings, caging and mailing her to the beekeeper either locally or overseas, ensuring that she arrives in the best possible condition at the end of her journey.

2.4.1 MARKING AND CLIPPING WINGS

A queen may be marked if she needs to be located regularly and quickly in queen rearing manipulations. By marking a queen it may be possible to determine if she has been superseded or, by using the international colour code, identifying what year she was produced.

To mark a queen, first find and remove her from the hive (section 2.2.5). If right handed, hold the queen by the thorax using the thumb and forefinger of the left hand. A number of queen catcher devices are available from beekeeping supply outlets if holding the queen is too difficult (see figure 2.2.6).

With the right hand, shake the bottle of marking fluid and then un-screw the cap (typists' correction fluid is suitable).

Wipe the surplus fluid from the brush and place a dab on the hairless dorsal (upper) surface of the thorax. Ensure that there is no fluid painted onto the antennae or the base of the wings or legs as this may affect perception and movement. Also, fluid running down the side of the thorax may block the spiracles (breathing holes).

Hold the queen firmly while the painted area dries. In the case where a numbered, coloured disc is to be attached, a small drop of glue is placed on the dorsal thorax and then a disc is dropped onto the glue and held down firmly until the glue hardens. This latter procedure is easily performed if the queen is anaesthetised with carbon dioxide.

If marking queens to record the year they were reared, use the international colour code with each year assigned a different colour on a five-year cycle: for years ending with the number 1 or 6 the queen should be marked white; 2 or 7 = yellow; 3 or 8 = red, 4 or 9 = green, 5 or 0 = blue. The discs and coloured marking fluid can be obtained from beekeeping equipment supply outlets.

In some cases the wings may need to be clipped. Wing clipping, as well as gluing a numbered disc to the thorax (figure 2.4.1), is often performed after instrumental insemination. The wings are clipped with a sharp pair of scissors with the fore and hind wing clipped on one side about halfway up the length. It is important not to clip the wing any lower, as there will be considerable loss of fluid from the costal vein.

Clipping may be used to record age with the left wing being clipped in odd-numbered years and the right wing in even-numbered years. Wing clipping does not necessarily prevent swarming, so a queen excluder should be inserted at the entrance.

2.4.2 QUEEN CANDY

Queen cage candy is made by mixing equal quantities of starch-free icing sugar with invert sugar such as fructose, glucose or dextrose and adding a few drops of water. The use of honey is not recommended as this may contain bacterial spores of American foulbrood.

Invert sugar is sweeter than sucrose hence more attractive to the bees. Fructose is preferred over glucose and sucrose as the latter two sugars form hard crystals, whereas fructose candy will remain soft and easier for bees to eat. Starch-free icing sugar is easier for the bees to digest than icing sugar with starch e.g. corn flour.

The invert sugar should be heated to 50-60°C to decrease viscosity and speed up mixing. The mixing should be done slowly with a few drops of water added at a time. The candy should be kneaded into a ball (figure 2.4.2) and left to stand overnight. If the ball loses shape overnight more icing sugar should be added. Store the candy in an airtight plastic container in a plastic bag to retain moisture.

The candy should not run when placed into the mailing cage as this will wet and possibly kill the queen. However, if the candy is too dry two problems may arise. The escort bees will not be able to liquefy the candy sufficiently from regurgitated honey stomach contents to consume it for energy and, second, when placed in the new hive, workers may not be able to release the queen by chewing a hole in the candy.

When using wooden cages it is advisable to coat the candy chamber in beeswax or paraffin wax. Some beekeepers cover the exposed candy under the mesh screen with a piece of waxed paper. Both methods will prevent the candy drying out.

Figure 2.4.1 Queen (centre) with numbered blue disc glued to the thorax.

Figure 2.4.2 Making up queen candy using invert sugar (dextrose), icing sugar and a few drops of water.

2.4.3 MAILING QUEEN BEES

The queen and a minimum of six escort workers are placed into mailing cages that have been provisioned with queen candy. The candy occupies about one third of the cage space.

The cages are then stacked on top of each other so that the screen faces are not in contact with each other.

The wire screen on the exposed face of the outer most wooden cage is covered by a piece of cardboard with a few small air holes punched in it.

The stack is then taped together with masking tape. The stack of cages can then either be placed in an envelope with holes punched in it and labelled 'live bees' on the front, or the stack can be placed into a plastic mesh bag with air holes and warning labels and the address stapled to the outside.

The queen breeder should advise the beekeeper when to expect the queen bees to arrive so they are not sitting in a hot mail room or letter box for a long period.

2.4.4 QUEEN INTRODUCTION

When the mailing cages arrive, separate them and place a drop of water onto the screen if it is warm weather. Keep the cages at room temperature in a cool dark place away from ants and fly-spray and place the mailing cages into hives within 24 hours.

Because it may be several days before the queen emerges from the mailing cage and begins egg-laying, greater success and less loss of queens results from introducing the cages into queenless nucleus hives. Once the queen has begun egg laying and her queen substance production increases, the nucleus can be united onto a queenless hive using the newspaper method (section 2.5.11).

If the caged queens are placed directly into strong queenless hives the queen may be superseded or even killed if there is a delay in egg-laying initiation, possibly due to a delay in queen substance production.

When introducing a queen in a plastic mailing cage remove the plastic tag from the candy end. For wooden mailing cages remove the cork from the candy end. Take a matchstick and push a small hole through the candy; this will assist in the nurse bees chewing through the candy. Place the cage between two brood frames (not honey

Figure 2.4.5 Banking frame for holding mailing cages to bank queen bees.

frames) in the queenless hive so that the air vents or screen are uppermost. The cage should be sloping so that the candy end is higher than the bee chamber to prevent dead escort bees from blocking the exit hole.

For the queen's guard mailing cages, slide the clear cover back to the release position so the candy is accessible to the nurse bees in the hive. Push a nail through the small hole at the back of the cover and suspend the cage between two brood frames.

Check the cages two days after introduction and if the queen is still inside open the cage and let her walk onto a brood frame. The nurse bees will have picked up the queen's pheromones through the cage vents by this stage and she should be accepted into the hive.

2.4.5 BANKING QUEEN BEES

When many queen bees are produced and caged ready to be transported to a beekeeper, or the queen rearer is preparing a large shipment of queen bees for export or for package bees, or the beekeeper receives a number of caged queens and has not prepared hives to receive them, it may be necessary to store these caged queens in a hive to care for them. This is called 'banking' queen bees.

Mated queen bees removed from mating nucs and placed into queen cages without worker escorts or queen candy can be placed with other similarly caged queens in a queen bank hive.

Many queen bees can be held until required in a queen bank hive. The individual cages are placed in a banking frame (figure 2.4.5). This may be a normal frame with a hardboard centre and 2-3 wooden shelves about 100 mm apart.

The cages are placed on the shelf with the wire screen or ventilation holes facing out. The cages may be held by a piece of elastic or a thin wooden slat that fits over two screws that extend out about 15 mm from each end bar. The banking frame is placed in the middle of either a queenless hive or a queenright hive, above a queen excluder. On either side of the banking frame are placed frames of emerging worker bees. These frames are replaced every five days with brood but not with bees.

When making up a queenless banking hive a number of different hives can be used and the cages are introduced at the time of making up the banking hive. It is important to provide young bees, as older workers may ball the queen i.e. clamp onto the queen and hold her tight. Wooden cages are preferred over plastic and the queens need to be able to get away from the workers so their feet are not damaged.

Other frames should be well provisioned with young brood and honey and the hive should be fed on sugar syrup and pollen substitute. The banking hive needs to be kept warm so the bees do not go into a winter cluster. It is possible to store up to 60 cages in one frame and have two frames (120 queens) in one banked hive. Queens may be banked for one to two months with few losses.

The banking hive should not have any fermenting honey or Nosema disease otherwise losses will increase significantly.

2.4.6 EXPORT CERTIFICATION

When exporting bees overseas, documentation must be completed to ensure that the queens being exported meet the health requirements of the importing country.

Government apicultural advisory officers can provide advice to beekeepers on the health requirements of the importing country.

In most cases the queen bee producer will be required to fill out an export health certificate based on the disease status of the hives from which the queen bees were produced. In some cases there may be a requirement for disease freedom within a certain radius of the apiary e.g. 5 km from the apiary of origin - and in some cases there may be the requirement for the country to be free of a particular disease.

A disease inspection may be carried out by an apicultural advisory officer prior to the endorsement of the health certificate. Failure to complete a health certificate correctly and meet the importation requirements may result in the consignment being destroyed when it reaches the importing country.

2.4.7 PACKAGE BEES

Package bee transit boxes hold about 1.5kg of bees, and one queen, and are supplied with queen candy or sugar syrup. The Arataki tube packages are long cylindrical plastic packages that are transported vertically with top ventilation. The tubes provide the bees with a greater degree of control over their environment, and reduced light penetration. The tubes take up half the space of conventional rectangular packages, are cheap, less labour intensive to fill and reduce bees escaping. Each tube has a leak proof feeding system made of a long sock of synthetic mesh filled with sugar syrup gelled with agar.   The packages arrive in prime condition despite lengthy journeys.

Package bees should be treated for Nosema. Fumagillin, an antibiotic, has been used extensively for controlling Nosema. Fumagillin, sold as Fumidil B, has a very long residual in honey with a withholding period of up to eight months; hence its use is only recommended for package bees and queen bee producers.

2.5 Swarming and Nucleus Hives

2.5.0 INTRODUCTION

Understanding swarming is crucial for managing hives in order to produce a surplus honey crop. Swarming is the natural way the bee colony reproduces. About half the bees in the colony leave with the old queen to establish a new colony some distance away (figure 2.5.0).

In New Zealand swarming usually occurs from late September to December when the numbers of bees in a colony are rapidly increasing. This rapid population increase means there is less queen substance being distributed to each hive bee. This decrease in pheromone levels per bee combined with overcrowding in the hive and other factors, stimulates workers to raise new queens in specially constructed swarm cells.

By understanding this phenomenon, the beekeeper can harness the swarming tendency to make up additional hives or strengthen weak hives. Swarming is also the best time for queen rearing as this is when the hives naturally rear their own queens. If the warning signs of swarming are not acted upon, however, the beekeeper may loose half the hive population.

2.5.1 PROBLEMS WITH SWARMING

Since the beekeeper can increase hive numbers by using artificial methods, the swarming of bees is not essential and is, in fact, a nuisance and undesirable because:

- swarming reduces the worker strength of a hive
- swarms often cannot be recovered
- lost swarms may actively compete with hived bees
- valuable queens can be lost in a swarm
- swarms may carry disease
- a hive that has swarmed may fail to gather a surplus honey crop.

Because of the disadvantages of swarming, prevention is better than cure. Thus, understanding the causes and indicators of swarming is essential so that steps can be taken to prevent or take advantage of this reproductive phenomenon.

2.5.2 FACTORS CAUSING SWARMING

There are a number of factors that have been suggested as contributing to swarming. As swarming is a complex process it is likely that no one factor alone is the cause but

Figure 2.5.0 Catching a swarm (top left) in a nucleus box.

that a number of factors operating together may give rise to a swarm. These factors would include:

1. An overcrowded hive with congestion of brood and increasing adult bee numbers in a confined area will result in conditions (high temperature) that encourage swarming. This is influenced by the size of the colony in relation to the volume of space available.
2. Colonies headed by an old queen (2 years old) are more likely to swarm than colonies headed by a young queen (1 year old). This is because the old queen is producing less queen substance. Also as the hive population expands the transmission of queen substance to all bees is reduced and hence the inhibition of queen cell production in workers is reduced.
3. A light nectar flow and an abundance of pollen, following a period when both were scarce.   This enables the bees to build up strength rapidly, with little nectar for them to gather.
4. Other factors also play a role such as the brood to bee ratio, worker brood rearing reaching a maximum level, an increase in younger age workers and external environmental conditions.

2.5.3 INDICATORS OF SWARMING

The building up of the worker population, the laying of drone eggs, the drawing of queen cups and the rearing of young larvae in them will result in swarming if the conditions are right.   These indicators should be looked for - especially the presence of developing swarm cells - and preparations made to avoid swarming. It is far easier to prevent swarming before it occurs than to attempt to capture the swarms after they leave the apiary. If the beekeeper can recognise the signs and take the necessary action to prevent bees reaching a frenzy known as 'swarming fever', then swarming can be avoided.

2.5.4 PREVENTING SWARMING

There are a number of techniques that the beekeeper can use to prevent swarming from occurring:

1. Artificial swarming (making up a nuc, top, split or division)

Artificial swarming is a very common method of preventing swarming. This involves removing sufficient frames of brood and honey from a hive that is close to swarming, to make up a nuc or division (section 2.5.5 and 2.5.8).

2. *Use non-swarming varieties/strains of bees*

Some races of bees, such as the Italian race, are less likely to swarm. Other races, such as the Carniolan's, are excessive swarmers.

3. *Keep colony strength below swarming level*

Reducing the strength of a colony by removing brood from strong colonies and introducing them into weaker colonies can prevent swarming. This means that strong colonies can be kept below swarming level, and their excess strength is not wasted. Ensure that all brood frames are checked for disease first before swapping frames.

4. *Replace old queens*

Old queens have a greater tendency to leave the hive with a swarm than do young queens. A regular requeening programme involving replacing the queens in half the hives every year is a normal aim for commercial apiarists, and replacing all queens annually is achievable by hobbyists.

5. *Give colonies ample room*

Providing space in the hive is achieved by supplying adequate supers prior to swarming, and introducing empty drawn frames into the brood area so that the queen has plenty of room to lay. If two brood boxes are used to over-winter hives, reverse the brood boxes during early spring disease checks so that the empty lower box is placed on top and the box that contains the bees, brood and honey is placed on the bottom. This rotation will delay swarming by three to four weeks.

6. *Destroy queen cells*

Destroying the queen cells the bees are raising when intending to swarm, will prevent them from swarming. Split the two brood boxes and tilt the upper brood box at a 45° angle. If no swarm cells are evident on the bottom of the brood frames in the upper box, the hive is unlikely to be preparing to swarm. If swarm cells are visible, all brood frames should be checked.

7. *Reorganise the brood*

Placing the queen in a new brood box on empty drawn combs with a queen excluder above this empty box and the remaining hive boxes above the brood chamber, will reduce swarming. The queen is forced to begin a new brood nest, as she is unable to get to the old one.

8. Move bees on to a honey flow

When a hive has a very large population of bees, with little nectar and ample pollen coming into the hive, there is a great tendency to swarm. By moving the hive to an area with a stronger nectar flow, much of the excess worker population will be redirected into gathering nectar.

9. Extract all honey from the hive

Removing excess honey should reduce brood raising, and encourage more foraging for nectar, thereby reducing the tendency to swarm. Extracted honey supers should be placed above a queen excluder so the queen does not lay in them.

10. Clip queen's wings

The clipping of a queen bee's wings will reduce the chance that the queen will attempt to leave the hive with a swarm. Very often however, the queen may attempt to fly and may be lost from the hive as she may crawl out of the entrance into the grass. This method of swarm control is not recommended. Placing a queen excluder over the entrance may be more effective. This method would usually only be attempted for inseminated queens.

11. Remove the queen

If the colony has no queen or larvae of a suitable age from which to raise a queen, they will not swarm. This measure should only be used as a last resort as it will weaken a colony.

2.5.5 MAKING SPLITS/DIVISIONS/TOPS

Artificial increases in the number of honey bee colonies can only be carried out when conditions favour such an increase. This is usually just prior to natural swarming during spring. Beginner beekeepers often make the mistake of trying to make rapid colony increases at the wrong time of year when conditions are unfavourable. The result is that the colony dies out and the beekeeper becomes discouraged. Beekeepers refer to making increases by the terms 'splitting a hive', 'dividing a hive' or 'making top hives'.

Some beekeepers use a Demaree board. This is similar to a division or split board except that rather than a small entrance (30 x10mm) the Demaree board has a swivel opening that can be opened or closed as required (see figure 2.1.7(b)).

To make up a division, or top, a full-depth box containing nine to ten frames of drawn out comb is required. Remove the drawn out frames and lean them against the hive nearing swarming. Place a division/split board (or Demaree board) on the upturned lid and place the empty box on top. Check the hive for disease. Remove two to three frames of sealed or preferably emerging brood and two frames of honey/pollen stores, with adhering bees and place these into the empty box. Do not separate the brood frames.

Check the frames carefully to ensure the queen has not been removed. Add additional empty frames to make the number up to nine. Place a single frame feeder into the box. Shake additional bees into the box from the parent hive if necessary. Replace frames removed from the parent hive with empty frames and ensure the parent hive has a feeder. Place the division board above the second brood box of the parent hive and place the new box (division/top) that has been made up, on top of the division board, with the new entrance for the division facing the rear of the hive. Introduce a ripe (ten day old) queen cell or caged queen between two brood frames in the top box. Feed both the parent hive and the top hive with 50:50 sugar: water.

In order to keep the bees from returning to the parent hive, some beekeepers block the entrance of the split board with newspaper or grass, or swivel the entrance closed on a Demaree board, for one to two days until the queen has emerged. Another option is to turn the entire hive 180° so that the original entrance is at the rear and the new division entrance is at the upper front of the hive. A sloping board up to the top entrance will assist workers to find the new entrance.

When the honey flow has begun and the queen in the top (division) has mated, the top hive can be united to the parent hive or used as the basis for a new colony.

2.5.6 TWO QUEEN HIVES

Two queen hives are made up by first producing a division or top hive above a normal hive. To produce a two-queen hive, once the queen in the top hive has mated remove the division board and replace this with a queen excluder. The two hives can then be united together with newspaper (section 2.5.11).

Workers will chew through the paper to unite the two colonies. Both queens will survive in boxes separated by the queen excluder, and the colony will function normally as a two-queen unit. The survival of the two queens will be enhanced if a pheromone excluder is used to separate the queens to minimise any fighting through the excluder. It is essential that a top entrance is provided for the top brood box to allow drones to exit the hive and not block the queen excluder. 'Supering' (adding boxes for gathering honey) will need to be twice the rate of a one-queen hive.

The two-queen colony can be left until late summer when the two queens, brood and bees can be separated to make two new colonies - or simply remove the excluder to revert to a one-queen hive. There is no need to remove one of the queens, as the younger queen usually kills the older queen.

2.5.7 NUCLEUS HIVES

A nucleus hive (nuc) contains a small colony usually with three to five frames of bees and brood. Nuclei are an essential part of modern beekeeping. They can be used as the starting point for a new, full strength colony or can be used as small colonies from which queens can take their mating flights. A nucleus hive may also give pleasure and confidence to the beginner beekeeper, who may be discouraged and put off beekeeping by having to handle large populous colonies.

2.5.8 MAKING UP A NUCLEUS HIVE

The nucleus hive is the same length and depth as a standard hive box, but is only half the width (section 2.1.8). Making up a nucleus hive requires a nucleus hive box, with four empty drawn out frames, a single frame feeder and a lid.

Before making up a nuc check the parent hive for American foulbrood.

When taking frames of brood from hives, care should be taken to ensure that the queen bee is not removed also. The best method is to locate the queen bee first and remove her, with escorts, using a queen's guard mailing cage. Once the required frames have been removed the queen can be returned back to her original hive.

A nucleus hive usually contains one to two frames of capped over or preferably emerging brood, one frame of solid capped honey, and one frame of honey and pollen. The honey and pollen frames are important as it may be some time before the young bees start to forage. A single frame feeder filled with dried bracken fern is usually included so that the hive may be fed sugar syrup. If a single frame feeder is not used then another frame of honey should be provided.

The frames for the nucleus can be taken from a single hive or from different disease-free hives. Young bees should be taken from one hive and left on the frames, as they will perform the normal hive duties. If there are few young bees on the frames, additional bees can be shaken in from brood frames of another hive. There should be enough worker bees to cover both sides of the four frames introduced into the nuc. Gauze should be placed over the hive entrance for one to two days to keep the workers confined, to allow the colony time to settle down, and to let the new queen emerge before worker foraging begins.

Frames removed from the parent hive should be replaced with drawn frames evenly staggered though the brood area.

The nucleus is then shifted 5km from where it was made up to prevent forager bees returning to the parent hive.

The nucs are arranged in pairs with entrances facing opposite directions. The nucs are then smoked and 50:50 sugar/water is added to the feeder (watering cans are handy for this purpose). Lastly, a ten-day-old queen cell is inserted into the middle of the brood area between two brood frames (not honey) and the nuc is closed up. Check after three days to see if the queen cell has hatched. If the queen has not emerged then introduce a new cell as soon as possible. Do not disturb the hive for two weeks after the queen has emerged, then check to see if she has mated. If there is no sign of eggs, leave the hive for a further one to two weeks, depending on the number of suitable days for mating, before rechecking.

Australian research by Rhodes and Somerville in 2003 suggests that the older the queen bee is when she is removed from the nucleus hive after mating, the higher the percentage likelihood that the queen will still be alive after 14 days and up to 15 weeks after introduction into a commercial honey producing hive. The research suggests that a queen should be at least 21 days of age before she is removed from the nucleus hive and that after 28 days in the nuc there is no additional benefit. In some seasons the long term benefits did not increase if the queen was left in the nuc after 21 days.

Alternatively a mated caged queen can be added. After two days, check to see if the mated queen has exited the cage and if not release her directly onto a brood frame by opening the cage.

Care should be taken with nucleus colonies as winter approaches. It is important to ensure they are well stocked with young bees and honey. A weak nucleus has little chance of surviving winter and should be united with a full size hive or made into a single nine frame hive with ample honey.

2.5.9 USES OF NUCLEUS HIVES

Nucleus colonies have many uses even though they do not produce surplus honey:

1. Requeening

Many commercial beekeepers maintain a number of nucs together in their 'out apiaries' (apiaries located away from the home base). If a production hive becomes queenless or the queen needs to be replaced, a nuc can be immediately united onto the production hive. Requeening by uniting a nuc to a parent hive has the highest acceptance rate for introduction of a new queen. This is because the queen is already laying and producing queen substance so there is little chance of supersedure. Also by having a nuc available for uniting, there is no loss of production from the parent hive.

2. Swarm control

Removing brood and bees from strong hives, to make up nucs during spring, will discourage hives from swarming by decreasing the hive population. If a colony has started to draw queen cells in preparation for swarming, it should be at an ideal strength for making up nucs. Remove two frames of emerging brood and two frames of honey and pollen.

3. Increase hive numbers

If a beekeeper wants to increase hive numbers, then prior to swarming, nucs can be split off expanding hives. If the nucs are placed in a new apiary site 5km away, they should build up rapidly. Once the nucs have within them mated queens, and are full of workers and food stores, they can be introduced into full-size hive boxes and frames of capped brood with honey added to speed up expansion.

4. Surplus queens

Surplus queens can be kept in nucs as a reservoir for emergencies. When no longer needed, the nucs can be united, built up into full-sized hives or added to weak hives.

5. Repair comb

Nucs will produce worker cells, even if not provided with foundation. Combs that have been damaged can be given to strong nucs to repair.

6. Maintain breeder queens

The life span of a queen selected for breeding will be extended if she is placed in a nuc where the demands for egg production are reduced. If the nuc is provided with a frame of empty drawn out worker cells she will lay in these cells and the developing larvae can be used for grafting.

7. Introduce shipped queens

Queen bees that are purchased from breeders and introduced directly into full-sized colonies are often superseded due to a poor initial egg laying performance. If the queen bee is placed in a nuc on arrival, she is given a chance to recover from the stress of being shipped, and can regain her egg laying capacity and queen substance production to such a level that it prevents the production of supersedure cells. Once the queen begins egg laying the nuc is united to the parent hive.

8. For beginners

Obtaining a three to four frame nucleus is an ideal way for the beginner beekeeper to develop confidence in handling bees that are relatively quiet and where there are not too many bees that may begin to sting. Once the novice develops confidence the nuc can be placed into a full-size box where it can expand naturally.

2.5.10 LOCATING NUCLEUS HIVES

Nucleus hives should be randomly located in a sheltered mating yard. This is due to the behaviour of the workers and the queen. Drifting occurs when bees become disorientated and return to the wrong hive. This may be due to factors such as strong winds prevailing at the entrance to the hive or bees having difficulty determining their own hive from another, especially if the hives are in rows or are painted the same colour. As a consequence a queen, returning to her hive from a mating flight, may be blown off course and drift into the wrong hive where she may be killed. The effects of workers drifting include: spread of disease, unequal strength hives, incorrect evaluation of potential breeder queens based on honey production, overcrowding resulting in swarming, an overall loss of up to 11kg per hive in honey production, hives at the end of rows and hives in front rows becoming very strong and the remaining hives becoming very weak.

To avoid such problems, nucs may be painted different colours or different shades of the same colour. Alternatively, different coloured symbols may be painted above the entrance or on the lid e.g. triangles, circles or squares may be painted, or silver metal shapes cut out and attached above the nuc entrance. Or, for top hives (splits) place the symbol or metal coloured disc (75 x 50mm) below the entrance. The order of sensitivity to colours for bees is: ultraviolet > blue-violet > green > yellow > blue-green > orange. Bees can also see a colour called 'bee purple' that results from the combination of ultraviolet and yellow, but cannot see red. Honey bees can distinguish between closed and open shapes and between unbroken and broken shapes (section 1.13).

Nucleus hives are often placed in pairs for stability against wind and animals, with their entrances facing in opposite directions to reduce drift (figure 2.5.10). If the nucs are touching each other, heat may also be transferred from one nuc to the other. If the nucs are placed in groups of four each entrance should be facing a different direction. The nucs can be arranged in different non-repetitive layouts including a circle or a 'U' shape with entrances facing inwards or outwards or alternating in and outwards. Other layout patterns include 'S' or 'W' shapes or four hives in a square (one hive at each corner) with all entrances facing in to the centre or outwards.

Making sure that there are landmarks such as trees and bushes in the apiary toward which the queen can orientate, is very important. Rocks, posts or any other structure that is a little different from the surrounding landscape can assist in orientation. The pairs of nucs should be 1-1.5 metres apart.

When locating a mating yard check that the site is not too wet or damp, and with good air drainage (air flow). The site should be protected from strong wind, have plenty of sun, be fenced from stock and be close to a source of water, nectar and pollen. Nucs should preferably be facing east through to west. Facing nucs to the cold southwest should be avoided unless good shelter is provided. The morning sun will warm East facing nucs and encourage foraging to begin earlier.

2.5.11 UNITING HIVES

Uniting of hives is the process of combining two hives into one. There are a number of reasons why a beekeeper may unite hives including (i) to join two weak colonies that on their own would not survive the winter or gather little honey (ii) to join a queenright nuc to a queenless colony (iii) to reunite divisions in colonies that have been made for swarm control (iv) to unite a swarm to a hive.

The best technique for uniting hives is the newspaper method, which if undertaken correctly, is virtually fool proof. As with making up a nuc, start by checking to see if there is any sign of American foulbrood before combining the two hives.

If both hives have queens, find and remove the less desirable queen and shake all the bees from the strongest hive into the bottom brood box. Place a sheet of newspaper on top of the bottom brood box and make a few slits in the newspaper with a hive tool (figure 2.5.11). Place an empty box on top of the newspaper. Introduce nine frames

Figure 2.5.10 Layout of nucleus hives using different arrangements to prevent the queen from drifting into the wrong nucleus: (top line left to right) pairs of nucs facing opposite directions supporting each other 1m apart; four nucs in a square facing different directions; in a cross with entrances facing inwards; in a circle facing out; in a cross with entrances facing out (bottom left to right) 'U' shape facing in and 'S' shape.

Figure 2.5.11 Uniting hive with newspaper, putting slits in newspaper to encourage bees to chew through when two colonies are united.

with bees and the new queen into the empty box, and replace the inner cover and lid. If the new hive originated from a nuc place the nuc frames in the centre of the top box then add four or five additional frames to make the number up to nine.

The bees from both hives will gradually chew through the newspaper and unite to form a single colony. By using the newspaper, the joining of the hives will be gradual and the bees will be too confused to fight. It is best to leave the united hive undisturbed for a week, and then re-organise the frames of brood, pollen and honey and remove any paper remains that may encourage waxmoth.

An alternative method of uniting two hives is to remove one queen, spray the bees from each hive with a fragrant aerosol spray and then combine the two hives together without newspaper. The spray confuses the bees so they will not fight.

2.5.12 SHIFTING NUCLEUS HIVES

Nucs can have a piece of gauze nailed over the entrance to allow air flow but prevent the bees from escaping. Alternatively metal discs with options of (i) air vent, (ii) queen excluder, (iii) open or (iv) closed, can be inserted over the entrance with the air vent option introduced when shifting. Paper can be used to block the entrance if the nucs are being moved short distances in cold weather - otherwise some ventilation should be provided during transportation to prevent overheating.

2.6 Honey Bee Nutrition

2.6.0 INTRODUCTION

Honey bees, like most animals, require, carbohydrate, protein, vitamins, minerals, fats (lipids) and water for normal growth, development, maintenance and reproduction.

Honey bees satisfy all their nutritional requirements by collecting nectar, pollen and water. By understanding the nutritional requirements of honey bees it is possible to make the best use of natural flora and correctly supplement the bees' diet by artificial feeding when the natural flora is deficient.

2.6.1 FOODS COLLECTED BY HONEY BEES

1) Nectar

Nectar and honeydew supply the carbohydrate requirements of honey bees. Nectar is collected from floral or extrafloral (outside the flower) nectaries. Nectar contains 5-80% sugar; water is the other major constituent with small amounts of minerals, vitamins, lipids, organic acids and volatile oils. Sucrose, fructose and glucose are the major sugars in nectar. Bees convert nectar to honey by evaporation of water and the addition of enzymes such as invertase which invert the disaccharide sucrose in nectar, to the monosaccharide sugars fructose and glucose in honey. Honey also has minor amounts of at least 22 other sugars. Honey is made up of about 80% sugar, 17% water and 35 other components including pollen, protein, vitamins, minerals, acids, enzymes and pigments.

2) Pollen

Pollen is the major source of protein, minerals, lipids and vitamins for honey bees. Protein is used for brood rearing including adult worker production of royal jelly and queen egg production.

The crude protein level for plant pollen varies from 7% for pollen collected mostly from pine trees to 37% if collected from agricultural areas. Fresh pollen contains protein (22% average), carbohydrate (sugars) 31%, lipids 5%, minerals 3%, water 11% and unknown 28%. Pollen is also rich in B-complex vitamins, enzymes, ascorbic acid, small amounts of starch and in rare cases may contain alkaloids or saponins that are toxic to bees. Pollen returned to the hive on the pollen baskets (corbiculae) of workers is treated with an acid from the brood-food glands to prevent germination and fermentation and when packed in the hive is known as 'bee bread'.

3) Water

Water is gathered by bees and used to dilute honey, provide minerals and maintain hydration in the blood and body tissues. The mineral content of water will vary depending on the source. Considerable amounts of water are also derived from nectar.

2.6.2 NUTRITIONAL REQUIREMENTS OF HONEY BEES

From nectar, pollen and water honey bees derive protein, carbohydrate, vitamins, minerals and fats, and together with water, this provides the nutritional requirements for honey bee growth and survival. A deficit in one or more of these components may, in the worst situation, result in colony death and at the very least result in a colony that performs below its potential.

1.   PROTEIN

Protein is derived mainly from pollen and is needed for all body tissue development and reproduction. Honey bees require pollen with at least 20% crude protein. Pollen contains from 7-37% protein. Pine tree pollen, with 7% crude protein, is unsuitable as a protein source whereas white clover, with 23-25% crude protein, would provide the minimum protein requirement for honey bees.

The level of body protein in honey bees may vary considerably from 21 to 67%. Management of hives should reduce extreme fluctuations in body protein and maintain a minimum level of 40% where possible. Worker bees in colonies with body protein levels below 40% have a life span of 20-26 days whereas workers with above 40% body protein live for 46-50 days.

Proteins are not simple compounds and when they are digested they break down into amino acids. If a protein contains an amino acid essential to a honey bee, it is said to be of high quality.   A protein of high quality for the honey bee, may not necessarily mean it is of high quality for another animal.

Building body tissue is a bit like building a house. The protein can be likened to general building materials while the amino acids are like the individual materials such as glass and nails. In digesting the protein, the honey bee can manufacture some of the required amino-acids from other amino acids that may be surplus, just as a builder may use a surplus stud to make a door frame. However just as with building materials there are some specific materials such as nails and glass that cannot be reconstituted from other items and these materials are required in certain minimum quantities.

Likewise there are some amino acids that cannot be reconstituted from other surplus amino acids. These are called essential amino acids.

De Groot in 1953 determined that there are ten essential amino acids required for normal honey bee growth and development, and these are required at levels ranging from 1- 4.5% (table 2.6.2).

If one of these essential amino acids is not present at the minimum level required, a bee cannot utilise all the protein they have digested. For example, if a bee consumed lucerne pollen with a total crude protein level of 20% but the level of iso-leucine was only 3% instead of the minimum level required of 4% then the bee would only be able to make use of three-quarters of the 20% of crude protein available (¾ x 20% = 15%) or 15% crude protein and pollen consumption would need to increase to compensate for this shortfall. An actively breeding hive needs 50-100 grams of pure protein per day to meet the protein needs of the hive. Hence if bees from a hive are gathering pollen from a flowering plant where the pollen contains 20% crude protein, and the hive requires 100 grams of pure protein per day, the bees will need to gather 500 grams of pollen per day to satisfy the colony protein needs. If the pollen was deficient in iso-leucine and only 15% of the crude protein was available the bees would need to collect an extra 166.7 grams per day to meet the pure protein demand for the hive.

2. CARBOHYDRATES

The chief sources of carbohydrates for honey bees are nectar and honeydew. Honey bees obtain carbohydrates from sugars derived from nectar, honey, honeydew, pollen and royal jelly. These sugars are used for providing energy for growth and reproduction. Collection of fresh dilute nectar stimulates bees to breed. The more fresh nectar entering the hive the more the queen is stimulated to lay eggs.

The sugars also provide energy for adult bees and are stored in the haemolymph. Adult honey bees can survive on carbohydrate alone but brood rearing requires protein, lipids, vitamins and minerals. Sugar consumption by worker bees is required for such functions as the production of royal jelly and beeswax. Approximately 7 kilograms of honey is consumed by young worker bees to produce 1 kilogram of beeswax.

3. WATER

Water makes up a high proportion of a bee's body. Its main function is to provide a medium in which chemical changes can occur. Bees without water will die in a few days. Bees use water for carrying dissolved food materials to all parts of the body,

Table 2.6.2. Essential amino acids for honey bees

Amino-acid	Minimum required percentage of amino acid in protein digested
Threonine	3.0
Valine	4.0
Methionine	1.5
Leucine	4.5
Iso-leucine	4.0
Phenylalanine	2.5
Lysine	3.0
Histidine	1.5
Arginine	3.0
Tryptophan	1.0

removing waste material, and digesting and deriving energy from food. Water is used for producing larval food, dissolving crystallised honey and regulating the temperature and humidity of a hive. Water is not stored but collected only when needed. Clean fresh water should be provided close (within 200m) to the hives especially during hot weather. Flotation devices e.g. corks, should be provided on top of water tanks to prevent bees drowning.

4.   FATS (LIPIDS)

Lipids are the most convenient form of stored energy and make up an essential component of the cells in the bee's body. Honeybees require lipids for growth, development including spinning a cocoon, reproduction and for producing moulting hormone. Cholesterol assists bees with brood rearing. Lipids are derived from pollen but are present at relatively low levels i.e. 4-8% (average). Bee foods with 6% vegetable oil are more palatable to bees than foods with lower levels.

5.   VITAMINS

Vitamins may play an important role in normal honey bee larval development and survival rates and in development of the adult hypopharyngeal glands involved in brood rearing. Vitamins are derived from pollen although this may vary depending on the nutritional value of the pollen. Generally, bees collecting a range of pollen sources will provide the colony with high levels of the vitamin B complex, vitamin C and carotene, the precursor of vitamin A. Pollen has no (or very low levels of) vitamin A, D, E and K. Excess levels of vitamins are toxic to honey bees.

6.   MINERALS

The role of minerals for honey bees is unknown. Minerals are derived mostly from pollen that contains most of the major minerals and is particularly high in potassium, phosphorous, calcium, magnesium and iron, and has lower levels of sodium. Phosphorous and potassium are the main minerals found in the body of honey bees. Water will provide a varying mineral content.

2.6.3 BEE METABOLISM

1.   Queen Bees

When queen bees first hatch, they eat honey directly from the comb, or are fed by worker bees. As queen bees develop they are fed a protein diet consisting of royal jelly and honey produced by older worker bees. This feeding occurs throughout the year.

It is important that queen bees receive a diet high in protein because:
-        the queen requires a protein diet for the development of her ovaries

- the production of eggs and pheromones is very demanding
- queen bees continue to grow after emergence
- queen bees live a lot longer than workers in unmanaged colonies (79% survive for 1 year, 26% for 2 years)

2.   Worker Bees

The first thing a worker bee does when it hatches is dry its wings and feed on honey and pollen. Twelve hours after emerging 50% of all workers will have eaten pollen. Pollen consumption reaches its maximum at five days of age and is required for growth of internal organs and gland development. Without this pollen the worker has a shorter life span and poor development (average life span is five to six weeks). Newly emerged workers are fed worker jelly by older nurse bees.

The increase in a bee's body weight is associated with the consumption of pollen and the development of the brood-food glands and fat bodies. Bees cease consuming pollen when they reach eight to ten days of age unless producing worker jelly for feeding larvae.

Most of the protein reserves of a bee are stored in their haemolymph or in their fat. From here it is used to produce worker jelly and repair tissues. The amino acids in pollen stimulate the activity of the brood-food glands. Insufficient protein from low grade pollen consumed early in adult life results in poor glandular development and hence a reduced ability of adult nurse bees to feed larvae, and a shortening of the adult life span.

Older bees that have finished nursing duties (10-14 days old) need only carbohydrate for energy derived from honey, and hence will die if honey is not available. All materials for repairs (vitamins and fats) are obtained from body stores originally derived from pollen.

It is important to ensure larvae are well fed as this directly influences the adult life span and ability to perform hive and foraging duties.

As bees age, instead of producing royal jelly, they convert the sugars in nectar into honey sugars. Bees have no storage organs for sugar. Sugars are stored in their haemolymph. The blood-sugar level is related to a bee's activity and changes rapidly. The higher the blood sugar level of the bee the greater is the activity that the bee may undertake. For example, worker bees that are foraging have a very high blood sugar level, while workers at rest have much lower blood sugar levels.

3.   Drones

The food intake of drone larvae is higher than that of workers at the larval stage. The diet of the adult drone is very similar to that of the worker bee. Drones do not forage for food. During the first few days of adult life they are fed entirely by workers and then gradually begin to feed themselves from honey cells. After a week they feed themselves entirely. Drones are fed by young workers on 'worker' jelly, pollen and honey. Older drones feed exclusively on honey. Drones need to increase their blood-sugar level for mating flights. They also need a high protein diet for the maturity of their reproductive organs.

2.6.4 PRACTICAL ASPECTS OF BEE NUTRITION

Queen rearing is mostly undertaken in spring and autumn because there are no drones available for mating in winter, and summer requeening interferes with honey production. In some areas natural flora provides adequate nutritional needs for queen raising, whereas in other areas or in certain seasons, artificial feeding is required to supplement the shortfall in the natural flora.

Natural Flora

Non-migratory beekeepers may plant nectar and pollen yielding plants close to their apiary sites, selecting plants that will supplement the existing pollen and nectar sources.

Migratory beekeepers are in a more fortunate position in that they are able to plan their moves in an attempt to ensure adequate food supplies for their bees.

Artificial Means

Bees may be fed on food supplies that cannot be obtained under natural circumstances. It may be necessary to only supplement the natural food supplies to a limited extent, or a beekeeper may have to substitute the entire food requirements of the colony. Practical feeding of bees uses two main food sources:
    1.   Carbohydrate (sugar) feeding;
    2.   Protein (pollen substitute or supplement) feeding.

2.6.5 CARBOHYDRATE FEEDING

Carbohydrate provides the major energy source for bees. Normally bees obtain carbohydrate from honey stores or freshly collected nectar. It is the freshly collected nectar that stimulates bees to breed. The larger the quantities of fresh nectar the more the stimulation for breeding. Honey stores are only used for maintenance during winter, and in maintenance of a breeding cycle.

Types of feeding
a) Honey
Frames of honey can be introduced into the brood boxes during queen rearing to replace empty frames. The frames should preferably be uncapped but if they are capped the cappings can be scraped off with a hive tool. This will make the honey more accessible to nurse bees to utilise for energy. If using honey from other hives ensure these hives have been checked for American foulbrood first.

b) Sugar syrup
Sugar syrup feeding simulates a light nectar flow and stimulates bees to breed. Syrup is fed to beehives using a frame, top, (figure 2.6.5) or Boardman sugar feeder.

Either white or raw sugar can be used to make syrup for feeding bees. Brown sugar is of no use, as it causes dysentery. At least one empty frame should be available for the bees to store the sugar syrup in the brood area.

To make syrup for stimulating the queen to lay or workers to feed royal jelly:
- Dissolve one part sugar into one part water (by weight).

Hot water will dissolve the sugar quickly. Once dissolved, the syrup should be left to stand. Feed to bees when the syrup is lukewarm. Feeding should begin two weeks before queen rearing to stimulate breeding and royal jelly production. Each hive should be fed a total of six litres in one feed or at least one to two litres per hive per week (figure 2.6.5). If the hives are being used for royal jelly production, discontinue feeding sugar syrup during the harvesting period to ensure the royal jelly contains only 'honey-derived' sugars.

This feeding regime will stimulate breeding just like a nectar flow would. This is an especially valuable method for stimulating queen bee rearing or to increase the hive population. If a maintenance feed is required to keep bees alive during autumn or spring then two parts sugar should be dissolved with one part water (by weight).

c) Dry raw sugar
Feeding dry raw sugar is a popular feeding method for keeping bees alive that have little honey. White sugar cannot be used as it hardens when it absorbs moisture. Dry raw sugar should only be used as a maintenance feed and not to rescue a starving

Figure 2.6.5 Sugar syrup feeding of a cell builder hive using a top feeder.

Figure 2.6.7 Feeding pollen substitute to cell raiser hive by smearing above and below the cell bars.

colony or to stimulate bees for queen bee rearing. Dry raw sugar is poured on to newspaper spread over the top of the brood chamber. It is important to ensure that there are a few holes in the newspaper to allow the bees to move through. Top feeders with two compartments, one for dry raw sugar and one for sugar syrup - are popular for dry raw sugar feeding with entry for workers via holes between the two compartments. To maintain a hive through spring 6kg of sugar is recommended.

2.6.6 PROTEIN FEEDING

Protein feeding is the feeding of hives either (i) with pollen, collected from pollen traps, refrigerated until required and mixed with a range of products, or (ii) feeding with pollen replaced by finely ground high protein vegetable meals e.g. soy flour, canola flour, brewers yeast or similar protein feed.

In the spring there is usually plenty of pollen although its protein content may be low. The feeding of soy flour or other protein flours as a supplement to pollen will increase the protein uptake of the hive, increase the hive population, and increase the hive strength before the start of queen bee rearing.

When bees are short of pollen for brood raising they will gather a variety of substances in a useless attempt to remedy this pollen shortage. Since pollen is essential for the development of queen larvae, and adult bees are short-lived without pollen, the beekeeper has to ensure that colonies are well supplied. If there is less than half a comb of stored pollen in the hive, then protein feeding is recommended, particularly if an ongoing programme of queen rearing is undertaken.

2.6.7 POLLEN SUPPLEMENTS

If a hive is collecting pollen, then there is no need to feed it pre-collected pollen. Pre-collected pollen is an expensive source of protein and has many difficulties associated with its use. However, it is a very important source of minerals and vitamins for honey bees. The addition of pollen will increase the palatability of vegetable protein flours.

Pollen can be collected from beehives by using pollen traps. It is very important to make sure that the hives from which pollen is to be collected are free from disease, as pollen is an ideal medium for the spread of honey bee diseases, especially American foulbrood and chalkbrood.

Protein feeding of bees can also be profitably undertaken during a low pollen but high honey floral source period. This occurs regularly when bees are foraging on some Australian eucalyptus where the pollen may be deficient in an essential amino acids. By providing the extra protein the vigour of the hive will be maintained, allowing it to continue rearing queen bees and gathering honey. Also the hive should continue to thrive after queen rearing is completed so that a surplus honey crop can be obtained.

The development of a satisfactory pollen supplement has been long and difficult but a suitable recipe now exists:

Pollen	1.50 kg
Lactalbumin or caseinates	3.28 kg
Brewers yeast	6.56 kg
Sugar	18.67 kg
Water	4.5 litres

A clean concrete mixer is the most efficient means of mixing the supplement. Pollen and 3.5 litres of hot water are mixed until the pollen has dissolved. The brewers yeast is added and then the lactalbumin is added very slowly. As the mixture thickens, more hot water is added until it achieves a firm consistency.

The mix is then formed into 500g patties, approximately 10mm thick. Each patty is stored until required between two sheets of wax paper in a freezer at -15°C, to kill wax moth.   The pollen patties are fed to the cell raising hives by placing between the frames rather than on top of the brood box or smearing the patty into cells of an empty drawn frame. This has been shown to give a 20-35% improvement in royal jelly yield, due in part to an increased queen cell acceptance and an increase in hypopharyngeal size of newly emerged worker bees.   Alternatively the patties can be rolled into a sausage shape and placed on top or beneath the cell bars (figure 2.6.7). Supplements, or substitutes, placed above or below the brood area will be consumed by forager bees rather than nurse bees, because workers move out of the brood area as they get older.   However, young nurse bees, consuming patties between the frames, will covert this to royal jelly and feed this to developing queen cells.   Direct transfer of food, in the form of royal jelly fed to older larvae by nurse bees, is less frequent when protein patties are placed above the brood frames where the queen cells are located, compared with providing protein patties within the brood area where the queen cells are located. The latter provides a more efficient nutrient flow to young and older queen larvae.

A recipe for a high protein and low carbohydrate supplement is as follows:
- 2 parts soy flour
- 1 part brewers or torula yeast
- 1 part disease free pollen

All the ingredients are mixed with a little disease free honey to form a mixture which has a crumble/putty consistency. It is fed dry between the frames inside the hive.

2.6.8 POLLEN SUBSTITUTES

The main difference between pollen substitutes and supplements is that substitutes have no pollen in their mix. Substitutes have the advantage of having no bee products that could potentially spread diseases such as American foulbrood. However, because of a lack of bee products, substitutes are often not as attractive and in some cases lack essential vitamins and minerals present in supplements. While supplements may have a higher nutritional value the disease risk may negate the advantages.

The most successful recipe for a substitute is the Beltsville bee diet developed by the United States Department of Agriculture consisting of:

        Lactalbumin or caseinates (120g)
        Deactivated brewers yeast   (230g)
        White sugar   (650g)
        Water to half the weight of sugar (320ml)

The ingredients are mixed into 500g (10mm thick) patties as for the supplements. Supplying 500g patties every six days for a month increases royal jelly production by 2.5g per hive per harvest.

Another example of a substitute for high protein and high carbohydrate is as follows:
-     3 parts soy flour
-     1 part brewers (or other) yeast
-     3 parts sugar.

This mixture can be fed dry inside the hive or moistened with a little water to make a paste.

Because most substitutes have a low (1-2%) oil content, the addition of small amounts of oil, in the form of vegetable oil e.g. soy oil, or alternatively cod liver or fish oil at 4% of the total mixture, is recommended. This will increase the palatability to the bees.

Pollen substitutes can also be purchased in a ready to use form.

CHAPTER THREE: QUEEN BEE BREEDING

3.1 Genetics and Reproduction

3.1.0 INTRODUCTION

In an effort to improve the characteristics of bee stock and increase production of hive products, beekeepers have developed controlled breeding programmes. In order to breed better quality stock, it is first necessary to have an understanding of bee genetics.

3.1.1 GENETICS

Genetics is the science of inheritance or the study of how characteristics are passed from one generation to the next, from parents to their offspring. Characteristics, like hair colour and eye colour, are inherited (passed from parent to offspring) via a substance called deoxyribonucleic acid or DNA. This is the genetic information necessary for the development and function of the bee. Every plant and animal cell contains DNA. It is formed into long, string-like objects called chromosomes. Chromosomes are found in pairs in the nucleus of each cell. At points along the chromosomes there are genes. These are the units of inheritance.

Genes can be thought of like office files, each file containing certain information. They are kept in order in a filing cabinet. The chromosome acts like a filing cabinet, keeping the genes in order.

Because the chromosomes occur in pairs, each gene also has a pair. Each half of the gene pair lives at exactly the same point (locus) on each chromosome. Genes may exist in alternate forms called alleles.

If both genes of a pair have the same information e.g. hair colour; black-black, then the individual who has those genes is referred to as homozygous for hair colour. If the genes carry different information e.g. black-blond, the individual is said to be heterozygous for hair colour.

When an individual is homozygous, there is no problem as to which gene is expressed. But with a heterozygous individual, it is more complex; one gene in the pair will be dominant over the other gene. When this dominant gene occurs, it will be 'expressed', suppressing the effect of the other gene, which is called a recessive gene. For a recessive gene to be expressed, it has to occur in a homozygous gene pair.

The expression of genes can be more complex than simply one gene being dominant over the other. The gene pair may interact, for example, when the gene that produces white eyes in bees and the gene that produces brown eyes are in the same bee; it has red eyes.

The same genes may function differently in drones, workers and queens and may even change when the bees move into a different environment.

3.1.2 REPRODUCTION

Reproduction is the process by which genetic information is passed from one generation to the next. This genetic information is contained within the DNA inside the nucleus of each cell.

Cells are classified as either body or reproductive cells. Body cells are involved with the structure and function of an animal, while reproductive cells are involved only with reproduction of the animal. These two types of cells replicate in different ways:

1   **Mitosis** – Body cells replicate, no gametes (egg or sperm) are formed. The process involves straight cell division. The new cells produced are identical to the original parent cell. This occurs during normal growth of the animal.

2   **Meiosis** – Reproductive cells divide to produce gametes. The female gametes are called eggs or ova and the male gametes are called sperm or spermatozoa. This is a reduction division resulting in only half the number of chromosomes (16 in honey bees) in the gametes compared to the original parent cell (32) (figure 3.1.2(a)).

During meiosis the individual chromosome pairs duplicate (figure 3.1.2(b)) then the homologous chromosome pairs come together with one pair coming from the mother of the queen (producing the egg) and the other pair coming from the father (of the queen). At this stage the paired chromosomes from the mother exchange segments with the corresponding chromosome pair from the father. This is known as crossing over, where segments of one chromosome from the mother are exchanged for segments from another chromosome from the father. The gene combinations change

Figure 3.1.2(a) Meiosis (formation of gametes) involves two cycles of division.
This is the process that occurs in the production of the egg. In the first division
(reduction division) the parent cell divides into two intermediate cells. In the second
division ('mitotic' division) the arms of the chromosomes (chromatids) are pulled
apart. Two homologous pairs (4 in total) of chromosomes are represented in the
diagram. (1) Chromosomes exist in long uncoiled strands. The two grey chromosomes
originate from the mother (of the queen producing the egg) and the two black
chromosomes originate from the father (of the queen) (2) DNA is replicated to form
two chromatids, each chromosome appears as a double chromatid (3) Homologous
pairs of chromosomes line up at the equator of the cell in a completely random
process called random assortment. Homologous pairs of chromosomes become
entangled called crossing over and exchange segments of chromatid (4) The
chromosomes move to opposite poles of the cell. The original cell splits into two
intermediate cells (5) The chromosomes in the intermediate cells align themselves at
the equator of the cell (6) A second ('mitotic') division occurs with the chromatid
strands separating to opposite poles (7) The cell divides forming new gametes with
half the number of chromosomes (haploid) as the original parent cell (diploid).
**(Redrawn from Allan, R. and Greenwood, T. 2001. Year 12 Biology: Student
resource and activity manual).**

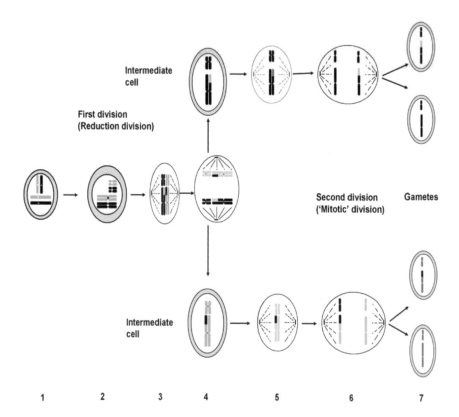

Intermediate cell

First division
(Reduction division)

Second division
('Mitotic' division)

Gametes

Intermediate cell

1   2   3   4   5   6   7

Figure 3.1.2(b) Meiosis involving crossing over of chromosomes during the formation of the egg (1) A pair of homologous chromosomes are represented in the diagram (the queen cell has 16 pairs or 32 chromosomes in total). The grey chromosome originated from the mother (of the queen) and the black chromosome originated from the father (of the queen). Four genes (A-D) are represented on the black chromosome and the recessive alleles (a-d) are represented on the grey chromosome (2) DNA is replicated to form two chromatids, each chromosome appears as a double chromatid (3) Homologous chromosomes align themselves at the equator of the cell and pair up (4) Chromatids of the homologous chromosomes become entangled, called crossing over (5) Parts of the chromatid of the chromosome are exchanged (6) Chromosomes undergo a reduction division and separate to intermediate cells (7) Chromosomes undergo a second 'mitotic' division and form gametes with each gamete having only one chromosome or half that of the original cell. Each gamete has a different combination of genes from the original mother (grey) and father (black), this gives rise to variation in each egg produced **(Redrawn from Allan, R. and Greenwood, T. 2001. Year 12 Biology: Student resource and activity manual).**

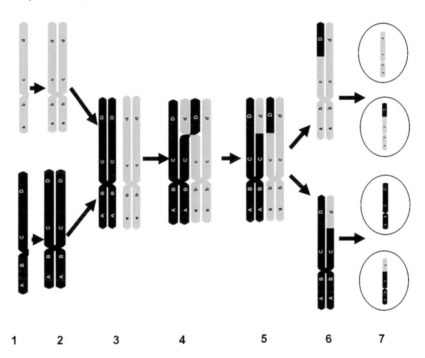

1	2	3	4	5	6	7

so that all possible combinations for the chromosomes from the mother of the queen and from the father of the queen will occur in different eggs laid by the queen.

The process of meiosis takes place in the ovarioles inside the ovaries of the queen during the production of eggs. During mating, the queen stores sperm from drones in the sperm sac or spermatheca in her abdomen. During the laying of the egg, if the egg is destined to become a female, sperm will be released from the spermatheca, move down the spermathecal duct and make contact with the egg as it is pressed against the spermathecal duct opening by the valve fold. The first sperm to enter the micropyle, a small opening at one end of the egg, will fuse with the egg nucleus. The male and female gametes (16 chromosomes each) will join together creating a zygote with a complete set of 32 chromosomes. This process is called fertilisation (figure 3.1.2(c)).

If the egg is destined to become a drone, sperm will not be released from the spermatheca, presumably the spermathecal valve remains closed (see figure 1.10) Sperm will not fertilise the egg, the unfertilised egg will be laid with only 16 chromosomes and will develop by parthenogenesis into a drone. However, most of the drone cells have more than 16 chromosomes due to doubling of the original set of chromosomes without division of the nucleus. This process is known as endomitosis and occurs particularly in insects, where doubling may be repeated many times in a single nucleus.

During development of female eggs after fertilisation, the cells inside the egg divide by mitosis, increasing in number. Eventually the egg hatches to a larva. Even though eggs that become drones are not fertilised, after egg laying, the cell also divides and eventually hatches to a larva. This process of cell division is called mitosis (figure 3.1.2(d)) and each new cell receives an exact copy of the parent cell's chromosomes.

Meiosis ensures there will be variation in the offspring. This is because fertilisation is a random process and you cannot predict which two gametes will join to form the new individual. However, in the case of bees, because all the sperm from a single drone are identical, variation is due to the genetic differences in the egg laid by the queen. The queen has a complete set of chromosomes (2 lots of 16 = 32). When she produces gametes (eggs) each egg is genetically varied due to the process of meiosis and crossing over of the chromosomes.

Figure 3.1.2(c) Fertilisation (formation of new individual) occurs when the sperm nucleus fuses with the egg nucleus. In the honey bee the sperm and the egg each have 16 chromosomes. After fertilisation the zygote that is formed has 16 homologous pairs or 32 chromosomes in total. One chromosome of each homologous pair in the zygote originates from the mother's (queen) egg (grey coloured chromosome) and one chromosome originates from the father's (drone) sperm (black coloured chromosome) (Note: the chromosome shapes used are a diagrammatic representation only).

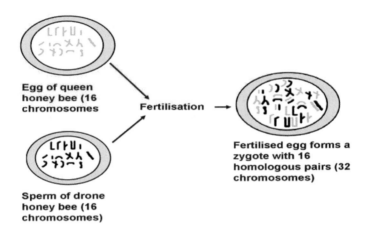

Egg of queen honey bee (16 chromosomes

Fertilisation

Sperm of drone honey bee (16 chromosomes)

Fertilised egg forms a zygote with 16 homologous pairs (32 chromosomes)

Figure 3.1.2(d) Mitosis (asexual reproduction) occurs during growth where two new 'daughter' cells are produced that are identical to the original parent cell
(1) The cell of a worker bee (female) has 32 chromosomes or 16 pairs (six are shown or 3 pairs) in the nucleus. The grey chromosomes originated from the mother's (queen) egg and the black chromosomes originated from the father's (drone) sperm (2) The nuclear membrane has disappeared, DNA is replicated to form two chromatids, each chromosome appears as a double chromatid and aligns along the 'equator' of the cell, with fine microtubules attached from the centre of the chromosome (centromere) to the 'polar region' of the cell (centrosome), the chromosomes split lengthwise pulling the chromatids apart (3) The chromosomes split in half, with each half moving to the opposite poles of the cell pulled by the microtubules (4) A new nuclear membrane forms around each cluster of chromosomes, the microtubules disappear and the cell divides into two identical daughter cells **(Redrawn from Laidlaw, H.H. and Page, R.E. 1997. Queen rearing and bee breeding).**

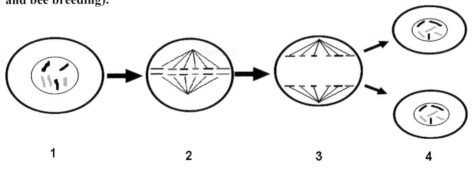

1   2   3   4

It is important to realise that the actual genes do not vary or change, but it is the way in which they are combined that changes. It is this changing of the gene combinations which provides the variation.

Both workers and queens develop from fertilised eggs. They have a full set of the parents' characteristics (32 chromosomes) and are referred to as diploid. The drone is different, because it develops from an unfertilised egg. Because of this the drone only has 16 chromosomes and is referred to as haploid. When the drone's gametes (sperm) are produced, each sperm gets a full copy of all the chromosomes, so there is no variation in the sperm from the drone. Hence all the sperm from a single drone are identical. Because the queen and worker are diploid and have two sets of genes, they can be expressing one genetic 'colour' and carrying genes for another colour. Whereas the drone being haploid has only one set of genes so it always expresses its genetic colour. For example, a yellow drone cannot carry the allele for a black drone.

A worker receives its genes from the queen and from one of the drones that the queen mated with. Variation still occurs even though the sperm of the drone is not varied. The variation in the worker comes from the genetic variation in each egg that the queen produces.

Drones produced in the hive receive their genes only from the queen. This is because the drones are produced from non-fertilised eggs. So it does not matter how many drones the queen mates with, as none of the genes from these drones are passed on to the sons i.e. the drones in the hive. The drones that mate with the queen pass their genes on to only their daughter offspring i.e. their daughters are the workers and new queens produced in the hive.

Another source of variation in the offspring of the queen is as a result of the queen mating with 12-17 drones during her mating flights. This results in her having a mixture of sperm in her spermatheca from each drone. The queen will pass on the genes from this mixed sperm in the fertilised eggs she produces that may become new workers or queens.

Variation is important to a species because it enables a species to adapt or change so that it can survive in a changing environment.

3.1.3 SEX ALLELES

When a queen lays fertilised eggs in worker cells, the eggs may develop into diploid workers or diploid drones. Normal drones are haploid and laid in drone cells. The sex of the fertilised egg laid in worker brood cells is determined at a single locus (position occupied by a gene on the chromosome) with multiple alleles (alternative forms of the gene). The number of alleles that determine the sex of the egg has been estimated to range from six to 18 in honey bees.

Individual eggs that are heterozygous (where genes have different information) at this locus become diploid workers and are raised as normal. Individual eggs that are homozygous (where genes have the same information) at this locus develop into diploid drones and are eaten by the workers soon after the egg hatches to a larva, possibly because of a substance secreted by the diploid drone larvae which causes worker cannibalism. Drones produced from diploid eggs are effectively infertile so would be a drain on the resources of the hive to rear them. Approximately eight variants of the sex alleles need to be maintained in a honey bee population to keep the brood viability above 90%.

In a closed population where there is no out-crossing, or where the queen mates with drones from a restricted gene pool, there may be a loss of sex alleles over a number of generations. With the loss of sex alleles there is an increasing risk that the queen will mate with drones with which she is related, resulting in inbreeding depression. This loss of sex alleles will eventually result in fertilised eggs produced that are homozygous at the sex gene locus and hence diploid drone eggs will be produced and eaten soon after hatching. In the worst case scenario up to 50% of the eggs laid may be diploid drones and because the colony is not producing enough diploid workers the colony will collapse and die. This can happen in a closed population breeding programme if no attention is made to ensure the queens are inseminated with a wide range of drone semen from unrelated breeding stock.

3.1.4 BREEDING FOR BEEKEEPERS

While it may seem complicated, beekeepers can still improve their stock using the principles of genetics. Instead of selecting individuals, as farmers do with sheep or cattle, beekeepers should select a whole colony. A superior colony should be chosen on the basis of brood rearing qualities of the queen and the behaviour of the workers.

The selected colonies should be evaluated over two seasons. During the first season the hive is evaluated alongside other hives in the out apiary using a yard sheet (section 3.2.3). The second season involves evaluating the top performing hives selected from the previous season in the one apiary using an individual queen record. By having all the superior colonies at one site, each hive will have an equal opportunity to perform. This makes hive comparisons more meaningful and selection of the best overall hives possible. The hives are arranged in an irregular pattern to eliminate environmental variables and this will reduce drifting. Providing that environmental variables are minimised the differences between hives, displayed in the individual queen record, should be mainly due to inherited genetic traits.

When a beekeeper selects a hive with outstanding qualities from the individual queen record it is important to realise that in order to reproduce a hive of a similar quality in future generations, the genes for the behaviour observed in the workers will be carried in the grandsons of that hive. In other words, to reproduce the outstanding hive in future generations, the breeder needs to produce daughter queen bees from that hive and the drones produced from these daughter queen bees will carry the genes of the outstanding hive. These drones should then be mated in isolation or using instrumental insemination with virgin queen bees reared from other unrelated, highly sought after hives with outstanding qualities. If the semen can be collected from as many of the grandsons (drones) of the outstanding hive as possible and inseminated into virgin daughter queen bees from other outstanding hives, more of the outstanding traits will be passed on in the offspring.

3.2 Stock Selection and Improvement

3.2.0 INTRODUCTION

The selection of breeding stock involves the collection of as much information about colony performance as possible and comparing this with colonies performing under similar conditions. It is important where possible, to base the selection on measurable characteristics rather than from a subjective assessment.

There are many characteristics which may be selected but most breeding programmes usually use honey production as one of the main selection criterion. The reasons for this are:

- most beekeepers want to make money and the more honey that is produced per hive the more money the beekeeper will make;
- honey production is measurable and can be easily recorded at the honey house;
- measuring honey production provides an opportunity to directly compare one hive with another in similar conditions and to make cautious comparisons between hive performance in different areas;
- production is often (but not always) a reflection of other traits e.g. queen egg laying capacity, rapid hive build up, disease resistance, willingness to enter supers through a queen excluder, and ability to draw comb.

A commercial beekeeper, operating 1000 hives and keeping good records, may be able to select the top 20-30 honey producing hives relatively quickly and from these hives further selection criteria can be used to evaluate the hives in more detail. During the second season it may be possible to select a limited number of breeder hives (5-10) for grafting queen cells and a further 10-15 for drone production.

3.2.1 RECORDING HONEY PRODUCTION

Recording the seasonal honey production of each hive is relatively straight forward providing that an accurate method of hive labelling is introduced (figure 3.2.1(a)). Each apiary must be allocated a letter, colour or symbol and each hive must be allocated a number. All the boxes on each hive, especially those removed for honey extraction, must be labelled with this number.

For example, the apiaries may be labelled 'A' to 'Z'; the hives in an apiary may be numbered 1 to 20, so on the first hive in apiary 'A' the number is A1 the second hive in apiary A is A2.

Figure 3.2.1(a) Preparing to take honey off to record honey production per hive, each apiary will be allocated a letter and each hive a number, all boxes from the one hive will have the same number eg A2.

Figure 3.2.1(b) Weighing honey boxes.

All the boxes on the first hive are numbered A1 so when they are removed it is easy to trace the origin of the box. This numbering system is also useful for disease purposes as a trace back. In the case of American foulbrood, where the disease is identified in a hive after the honey supers have been removed, all the boxes originating from that hive can be located and destroyed.

When removing honey from hives it is important to be consistent. For example if two full depth brood boxes are normally left on for winter then leave two full depth boxes on for every hive. Do not leave only one full depth on one hive and three on another. All hives should be left in the same condition at the end of the season and the surplus honey (i.e. above the excluder), removed to the honey shed for recording and extraction. Likewise, if the hives are to be fed before winter, they should all be fed the same amount.

When the boxes are ready for extraction prepare a chart for recording weights of supers. Each apiary site should be recorded on a separate page. The hive number should be recorded for each apiary. Using a set of scales weigh the full weight of each super, record the full weight alongside the hive number, and the number of frames per super (figure 3.2.1(b)). Remember that there may be 2-6 boxes per hive depending on whether full, three quarter, or half depth boxes are used.

The full weight of all supers from each hive should be recorded in separate columns under the appropriate hive number. Now scrape the top bar of each frame and write on it the hive and apiary number (e.g. A15). Use different coloured markers for different boxes and frames from the same hive, or a similar system.

The honey is then uncapped and extracted. The frames are placed back into the super of origin ensuring that the correct number of frames is returned. This also assists with preventing the spread of American foulbrood between hives. The box with sticky (wets) frames is then re-weighed. The empty weight is recorded for the box alongside the appropriate hive number.

3.2.2 ASSESSING HONEY PRODUCTION

At the end of extraction, data on full and empty weights of each super from each hive, for all hives from all apiaries, should have been obtained. Either using a computer program such as Excel or Quattro, or manually using a calculator, subtract the empty weight from the full weight of each super to obtain the amount of honey and wax removed. Add all the values for each super from one hive together e.g. hive A12 may have had 3 supers removed: the first super had 15kg of honey, the second super had

12 kg of honey, and the third super had 8 kg of honey so the total honey production of hive A12 is 15+12+8 = 35 kg. Add up all the production figures for all the hives from all apiaries. Now rank the hives from an apiary site according to their production per hive e.g. A16=45 kg, 1st; A7=42 kg, 2nd; A2=38 kg, 3rd.

Check to ensure there are no unusual figures, for example a hive that produced no honey. Was the hive on a site throughout the season? Or a hive that produced 60 kg, is this figure correct or was a mistake made? After checking the figures calculate the average production for an apiary by adding up production figures for each hive in the apiary and dividing by the number of hives in the apiary.

Example A1 = 22, A2 = 18, A3 = 32, A4 = 28.

$$\frac{22+18+32+28}{4} = 25 \text{ kg (the average for 4 hives at site A)}$$

In order to compare production figures between apiary sites it is necessary to calculate the percentage increase or decrease in production of each hive relative to the average production for that site and then compare the percentage increase/decrease relative to the average between different apiary sites.   This is because of the different environmental conditions at the different apiary sites.   An example will help to illustrate this comparison. Using the above figures from apiary 'A' the percentage increase or decrease can be calculated from the average of 25kg for 'A' site. Note that the average for 'A' site is 100%

Hive				
A1	=	22/25 (average for A site) x 100/1 =88% of average		
A2	=	18/25 x 100/1	=	72% of average
A3	=	32/25 x 100/1	=	128% of average
A4	=	28/25 x 100/1	=	112% of average

Now for apiary site B. The figures are B1 = 9, B2 = 25, B3 = 27, B4 = 18, B5 = 21.

The average for B site is:

$$\frac{9+25+27+18+21}{5} = 20 \text{ kg per hive}$$

The percentage increase or decrease of the average for the hives at B site is:

B1	9/20 x 100/1	= 45%
B2	25/20 x 100/1	= 125%
B3	27/20 x 100/1	= 135%
B4	18/20 x 100/1	= 90%
B5	21/20 x 100/1	= 105%

In order to compare the honey production figures between apiary A and apiary B check the percentage increases at each site. In the examples given hive A3 was the top-performing hive at 'A' site with 128% of the average for that site and this compared favourably with B2 (125% of the average for B site) and B3 (135% of the average for B site). Thus, if hives were to be selected for honey production and for further evaluation as possible breeders, from sites A and B together, hives B3, A3 and B2 should be chosen in that order from sites A and B.

3.2.3 RECORD KEEPING

When assessing hives for different traits two types of records can be made. The first is called a 'yard sheet' (table 3.2.3(a)). The yard sheet has the characteristics to be assessed down the left column and the hive number along the top row. Each trait is graded either on a 1 to 4 or a 1 to 5 scale e.g. 1 = unsatisfactory, 2 = satisfactory, 3 = good, 4 = exceptional, or 1 = poor, 2 = fair, 3 = good, 4 = very good, 5 = excellent.

When the score for each trait has been entered, total up the scores for each hive, then if a selection is to be made select hives with the highest scores. If two hives are similar in total score select the hive with the highest drone score. The second record is the 'individual queen record' (table 3.2.3(b)). This is a record of the colony assessed over a season and includes production information, parents if known, the marking of the queen and the same traits as in the yard sheet. The traits assessed are listed in the left column and the dates assessed are listed across the top row.

Based on the information built up over an extended period, queen mother and drone mother hives would be selected for breeding and artificial insemination. In order to avoid any inbreeding and loss of sex alleles, at least five and up to ten queen mother hives should be selected, and at least ten to fifteen drone mother hives should be selected as parents of the next generation.

Table 3.2.3(a) Yard (apiary) sheet (an example of a record of assessment for the top four hives from apiary A). See key below for details on scoring a two storey hive.

Hive number	A6	A10	A15	A18
Date of assessment	22-3-06	22-3-06	22-3-06	22-3-06
Honey production - average for apiary site =30 kg -kg of honey extracted from hive -% increase/decrease from average - score	 41 kg 137 % 4	 38 kg 127 % 4	 46 kg 153 % 5	 32 kg 107 % 3
Drone colour	5	4	4	3
Worker uniform colour	4	5	4	3
Brood viability (using rhombus)	4	4	5	2
Temperament of workers	5	4	4	4
Disease resistance -identify disease (eg sacbrood, chalkbrood)	5	4 1 cell sacbrood	5	3 2-3 cells chalkbrood
Swarming tendency	5	4	5	5
Population of workers	5	3	4	4
Honey stores	4	4	5	4
Pollen stores	4	5	4	3
Queen -comments on size and behaviour	Plump marked queen	Long	Plump quiet on comb	Not sighted
General comments on hive	Good hive Good drone colour	Average hive	Good hive Good honey production	Average may need requeening
Total score	45	41	45	34
Rank	1st =	3rd	1st =	4th

Key:

General	**5=excellent, 4=very good, 3=good, 2=fair, 1=poor**
Honey production:	5=150+%; 4=111-149%; 3=90-110%; 2=70-90%; 1=69% or less
Drone colour: (to select Italians)	5= 90-100% yellow; 4=60-89% yellow; 3=40-59% yellow; 2=11-39% yellow; 1=1-10% yellow (90-100% black)
Worker uniform colour:	5=all similar markings; 3=some variation 1=all variable
Brood viability:	5=95-100%; 4=90-94%; 3=85-89%; 2=80-84%; 1=79% or less
Temperament:	5=no movement on comb; 4=no stinging, quiet; 3=streaming on comb; 2=a few stings during inspection; 1=beating against the veil
Disease resistance (chalkbrood, sacbrood)	5=no disease; 4=one diseased cell; 3=2-5 diseased cells; 2= many diseased cells in one frame; 1=several frames with diseased cells
Swarming	5=no swarm cells; 4=one swarm cell; 3=2-3 swarm cells; 2=many swarm cells; 1=hive has swarmed
Worker population	5=very strong (all frames covered in bees); 4=strong (outside two frames empty); 3=average (3-4 frames empty); 2=50% of frames covered, 1=nearly dead
Honey stores	5=full box of honey (9-10 full depth frames); 4= 7-8 frames of honey; 3=5-6 frames of honey; 2=3-4 frames of honey; 1-2 frames of honey
Pollen stores	5=3 or more frames full of pollen; 4=2 frames of pollen; 3=1 frame of pollen; 2=½ frame of pollen; 1=no pollen

Table 3.2.3(b) Individual queen record sheet (example of record for A6 hive selected from yard sheet). The data taken from the yard sheet make up the first entry (22-3-06) for the individual queen record. Note the total score for the first entry does not include a score for the hygienic behaviour test.

New hive number (original hive)	Z1 (A6)
Honey production -kg of honey extracted from hive -% increase/decrease from average - average for apiary site =30 kg - score	41 kg 137 % 4
Marking (colour and number)	Blue 14
AI queen (wing clipped)	Left wing

Date of each assessment	22-3-06 (autumn)	21-9-06 (early spring)	26-10-06 (mid spring)	23-11-06 (late spring)
Honey production (previous season)	4	4	4	4
Drone colour	5	5	5	5
Worker uniform colour	4	4	4	5
Brood viability (using rhombus)	4	3	4	4
Temperament of workers	5	4	5	5
Disease resistance -identify disease (eg sacbrood, chalkbrood)	5	4 (2 cells sacbrood)	4 (3 cells chalkbood)	5
Swarming tendency	5	5	4	4
Population of workers	5	3	4	5
Honey stores	4	3	3	4
Pollen stores	4	3	4	4
Hygienic behaviour	Not recorded	4	4	4
Queen -comments on size and behaviour	Plump marked queen	Queen laid 3 frames of brood	5 frames of brood	9 frames of brood
General comments on hive	Good hive Good drone colour	Average hive low on food stores	Average hive fresh nectar	Good hive good population
Total score	45	42	45	49

Key: see also 3.2.3(a)

Hygienic behaviour   5=100% of cells uncapped and removed in 2 days
4= 75-99% cells uncapped and removed over 3 days
3= 50-74% cells uncapped and removed over 3 days
2= 25-49% cells uncapped and removed over 3 days
1= 0-24% cells uncapped and removed over 3 days

Honey production   Score taken from previous seasons total honey yield, this score remains constant for each entry

3.2.4 SELECTION CRITERIA

Selecting hives for breeding is made easier if honey production evaluation has been completed.   By calculating the hives with the highest percentage increase in production the top 2-5% of the hives from all apiaries can be identified depending on how many hives the breeder wants to further evaluate. This is normally undertaken in autumn at the end of the first season. By undertaking an inspection of these top performing hives with a list of selection criteria on a yard sheet, the top queen bees can be selected and returned to a home apiary for further evaluation. For example from 1,000 hives select 30 hives with the highest percentage production increase, evaluate using a yard sheet and return 20 of these hives to the home apiary for further evaluation.

With a copy of the yard sheet at hand the 30 hives with the highest percentage increase from various apiaries can now be evaluated using the characteristics below.

Drone Colour: The queen bee produces sons (drones) from non-fertilised eggs therefore the drones reflect the mating habits of the queen's parents or the grandparents of the drones in the hive. If the drones in the hive are varied in colour e.g. showing Italian and Dark honey bee characteristics, this suggests that the queen's mother may have mated with drones of a different race to herself. The problem here is that because of the racial differences in the grandparents it would be very difficult to reproduce the characteristics of the hive being evaluated in the queen bee's offspring (daughter queens).

If the grandmother of the drones in the hive (queen's mother) was Italian and mated with drones of the same race this is known as a pure bred mating and would likely be reflected in a more uniform colour from the drones in the hive. If the grandmother mated with Italian drones the drone colour in the hive should be predominantly yellow. Queen's produced from pure bred mated grandparents are more likely to produce uniform characteristics in their offspring similar to their mother's hive being evaluated.

Worker Colour: Worker colour reflects the mating habits of the queen bee currently heading the hive. If there is considerable variation in the abdominal colour marking of the workers this reflects the variation in the genetic make-up of the drones the queen bee mated with. An even colour marking in the workers reflects a similar genetic make-up in the drones sired to the queen and therefore any female offspring produced from this queen should closely resemble that found in the hive being assessed.

Brood Pattern: The brood pattern of the queen is determined by the age and egg laying capacity of the queen and the viability of the eggs. The viability of the eggs depends on the extent of inbreeding between the queen and the drones with whom she mated. The loss of egg viability is due to the loss of the sex genes or sex alleles. This occurs when the sex alleles in some of the fertilised eggs are alike. This situation arises when the queen mates with drones with which she is closely related. When the sex alleles are alike the larva is said to be homozygous for the sex alleles, and the larva that hatches from the fertilised egg is male and diploid instead of the normal diploid female. The larva is eaten by the worker bees, leaving an empty cell and a speckled brood pattern. When the sex alleles are different, the larva is female and is said to be heterozygous.

In order to assess egg viability young queens should be chosen. Using a piece of cardboard cut out the shape of a rhombus to cover exactly 100 brood cells. This can be undertaken by taking a piece of card about 80mm wide by 130mm long. Place the long end of the card alongside the cells on the brood and measure ten cells from outside wall to outside wall. This should be about 54 mm. Draw this line about 10mm in from the outside of the card but parallel to the long edge. Using a protractor at the left hand end-point of the line, measure an angle of 120 degrees and draw a line 54 mm long. At the right hand end point measure a 60° angle and draw a line 54mm long parallel to the first line.

Join these two lines with a fourth line to make a rhombus. Cut the inside of the rhombus out (figure 3.2.4(a)) and place the remaining card over the brood to see if it covers exactly 100 cells.

By moving the rhombus over the brood select a patch of near perfect comb where the larvae are one to two days old. Place the rhombus over 100 cells where all the larvae are exactly the same size and hence nearly the same age. Then count the cells in the 100-cell area that are NOT occupied with that age larvae. Subtract also, those empty cells and cells with pollen or honey or cells with un-emerged bees from 100. This gives the figure of the queen's viability. For example if you count 3 empty cells, 3 cells with pollen and two with honey within the rhombus then the brood viability is 3+3+2=8, 100-8=92% brood viability. Assessing viability using sealed brood is less accurate because as soon as an egg fails and is eaten the queen will re-lay in the cell so the sealed brood often represents pupae of different ages. If only sealed brood is present however, this may be a useful measure for comparisons between hives. Select hives with 95% brood viability or higher if possible (figure 3.2.4(b)).

Figure 3.2.4(a) Rhombus used for measuring queen egg viability.

Figure 3.2.4(b) Frame showing high (100%) brood viability of queen.

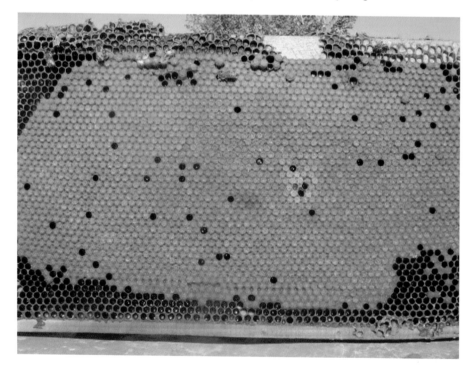

Disease Resistance: Where possible the hives should be free of disease. Never select a hive if there is any possibility of it having American foulbrood as this disease could spread to other valuable breeding stock with consequent destruction of all breeding stock. Where possible the hive should have very few or no diseases or disorders such as chalkbrood, sacbrood or halfmoon syndrome (disorder). A few cells (1-3) are probably acceptable. A record should be made of the number of cells infected and the type of disease.

Varroa treatments (figure 3.2.4(c)) should be introduced before brood disease assessments are made otherwise the results may be influenced by Parasitic Mite Syndrome (PMS). Note: formic acid treatment should not be applied to colonies used for drone rearing as this causes drone eggs to be removed from the comb and reduces the survival of adult drones with about half as many drones reaching ten days old compared with untreated colonies.

The ability of bees to detect and remove diseased or dead larvae or pupae from brood is known as cleaning or hygienic behaviour and assists in preventing the formation and accumulation of disease spores. The hygienic behaviour is under the control of at least two behavioural activities, one for detecting and uncapping dead or diseased brood and the other for removal of dead or diseased brood. The uncapping and removal behaviours are under the control of two independent genetic mechanisms so a worker may display one behaviour but not the other. Both hygienic behaviours need to be present in the worker population for the hive to effectively display disease resistance.

There are several ways to test for the presence of both cleaning behaviours. One method is to place a rhombus over 100 cells of sealed brood; however a small circle of seven cells is sufficient. Mark the wooden frame with coloured pins on the top and side bar so the exact location of the brood can be identified at a later date. Using a fine needle pierce each cell and push the needle into the base of the cell and carefully remove it. Make as small a hole as possible in the capping. An alternative method is to cut out an area of brood e.g. 100 cells, count any cells with pollen, honey, or unsealed brood, freeze the brood for 12-24 hours then place this back into the brood frame. A third method is to cut the top and bottom out of a round (75mm diameter) tin can. Place the can over a section of even capped brood and pour 100ml of liquid nitrogen into the can. The liquid nitrogen will kill the brood and evaporate. Mark the frame with coloured pins.

Figure 3.2.4(c) Checking sticky board for varroa mites 24 hours after placing in the hive. Varroa strips are placed in the hive between the frames with a sticky board placed on the bottom board of the hive.

With any of the above methods check the brood after one, two and three days and record the number of cells, either pricked or frozen, that have been uncapped, and the number that are completely clean. Colonies with both behaviours present should uncap and remove all dead brood in the test area within two to three days and some hives may perform this within 24 hours. If only the uncapping behaviour is present cells may be uncapped but dead larvae may not be removed. Hives removing 95% of the dead brood in 48 hours have sufficient hygienic behaviour to demonstrate varroa tolerance.

Suppressed Mite Reproduction (SMR): The number of female varroa mites that enter brood cells and produce offspring varies considerably. The number of female varroa mites that do not produce viable offspring is called the infertility rate. This rate is around 10-20% but may be as high as 40% in summer. To check hives for infertility, or SMR, check 20 mite infested worker pupae at the purple eye stage. Mature female varroa are oval and reddy brown, males are smaller and white to tan in colour, and the immature stages are white. Mite infertility is judged as either a single live or dead 'reddy-brown' female with no offspring. Percentage infertility is then compared between potential breeder hives.

Temperament: This can be assessed in two ways. Firstly, the degree of stinging of workers and flightiness, and secondly the amount of streaming or movement of workers on the comb. A poor hive is one where bees are hammering against the veil as you take the lid off. An ideal hive is one where using little smoke, a brood frame is removed and the bees remain calm even when you pass your hand quickly over the frame a few centimetres above the bees.

Swarming: The likelihood of a hive swarming is difficult to assess unless it is during the spring period or unless the colony is checked throughout the season. Ideally a colony should not swarm even when crowded, provided the queen is not more than one year old. If swarm cells are present along the bottom of the upper brood frame in the hive this should be recorded. If a hive has swarmed it should not be selected for breeding.

Supersedure: As with swarming, if supersedure cells are present on the face of the comb this should be recorded. Ideally the queen should not be superseded during the first two years of age. However, many average queens are superseded between the first and second year. It is important that the queen in the hive is the mother of the workers; if there is any doubt the hive should not be selected.

Queen Appearance: If the queen is located her behaviour and size should be noted. Ideally the queen should be plump with a wide thorax and a long tapering, symmetrical abdomen. All her legs and antennae should be intact. The size of the queen will not influence the behaviour of her offspring but may determine her ability to produce offspring. The behaviour of the queen should be calm and quiet and she should be searching for empty cells to lay in. If the queen and hive is to be selected for further evaluation, mark the queen with typist correction fluid or glue a numbered disc to the top of the thorax.

Other traits to consider: These would include long life of the queen, frames of bees, amount of brood, worker ability to draw foundation, ability to enter supers, ability to build up in spring, wintering ability, ability to collect and store pollen (particularly important for hives involved in pollination), absence of brace and burr comb.

3.3 Breeding Programmes

3.3.0 INTRODUCTION

The European (Western) honey bee, *Apis mellifera* (Linnaeus), belongs to the hymenoptera order that includes ants, bees and wasps and the Apidae family of bees that make nests and store pollen and nectar to feed their larvae. Honey bees belong to the genus Apis. Other types of bees may collect and store honey, but are not as useful for intensive production as the honey bee. The genus 'Apis' means bee and the species 'mellifera' means honey gatherer.

3.3.1 RACES OF HONEY BEE

There are approximately 24 different sub-species or races of the honey bee worldwide. These races are somewhat isolated geographically and are the result of natural selection in their homeland rather than breeding by beekeepers. All races have the potential to inter-breed, or to form hybrids if they are in the same area.

These races are naturally distributed through Europe, the Middle East and Africa. *Apis mellifera is* not native to Asia, America or Australasia.

There are four major races of honey bee that are of economic importance. They all originated from Europe.

a) Italian Honey Bee *(Apis mellifera ligustica* (Spinola))

The Italian honey bee originated from Italy and is the most commonly used bee world-wide and in New Zealand. They are smaller and lighter coloured than other honey bee races, with yellow bands on their bodies and a long tongue. Italian honey bees are generally gentle and are very good breeders, starting early in the spring and continuing through to the end of autumn. Italian bees are less inclined to swarm than most other sub-species but are inclined to rob other hives and consume large quantities of honey over winter. When conditions are good, Italian honey bee colonies have high populations and are very good honey producers.

b) Dark European Honey Bee (*Apis mellifera mellifera* (Linnaeus))

Dark bees are predominantly black on the abdomen and may have yellow spots on the second tergite; they have the largest body and longest abdominal overhairs of the European races with the shortest tongue. Dark bees produce lower volumes of honey and are more aggressive than the other three sub-species of bee. Dark bees survive the winter (over-winter) well due to their better ability to maintain heat within the cluster and greater resistance to Nosema and dysentery. Dark bees have a longer broodless period in winter and are slow to start producing brood in the spring; they consequently consume less honey stores in winter than Italian bees.

c) Carniolan Honey Bee (*Apis mellifera carnica* (Pollmann))

Carniolan bees originated from Austria, former Yugoslavia, Rumania, Bulgaria and Hungry. They are grey-brown to black in colour and similar in size to the Italian bees. Carniolans are very docile; they over-winter in small colonies and consume low quantities of honey through winter, and so are better suited to a harsh climate. They start brood rearing very quickly in the spring when there are good pollen sources and have a strong tendency to swarm. Carniolan bees are good honey producers and are not inclined to rob other hives.

d) Caucasian Honey Bee (*Apis mellifera caucasica* (Gorbachev))

Caucasian bees came originally from the Caucasus Mountains of southern Russia. They are similar to Carniolan bees in looks and are dark grey to black in colour. They are docile bees, with the longest tongues of the four races, which enable them to pollinate a wide range of flowers. Caucasian bees do not swarm and are greater users of propolis than other bees. They may be susceptible to Nosema disease. Caucasian bees are good honey producers and do not drift into other hives.

The commercial bee population in New Zealand is made up of a mixture of these four races, with the Italian race being the most widely used and the Dark race being predominant in native bush areas of New Zealand. Carniolan semen has recently (2004) been introduced into New Zealand.

3.3.2 IDENTIFICATION OF RACE

The different races of the honey bee are distinguished by up to 40 morphological (external form) features but a few of these features such as wing venation, and abdominal body hair and colouring, can provide a relatively accurate indication of which of the four major races are present. The main morphological features used to measure race include:

1) Cubital index
2) Discoidal shift
3) Abdominal overhairs
4) Abdominal tomenta
5) Abdominal body colour
6) Length of the proboscis (tongue)

1) Cubital index: All *Apis mellifera* honey bees have a similar pattern of wing venation but racial differences arise in the length and angles of some wing veins. On the front of the forewing there is a long costal vein. Behind this at the tip of the wing is the radial cell and below this are three cubital cells. There are two short veins in the third cubital cell marked 'A' and 'B' in the diagram (figure 3.3.2(a)). The ratio of the length of 'A' relative to the length of 'B' gives the cubital index. Another method of measuring the cubital index is to use a Herold fan. The wing is projected onto a wall and the Herold fan is placed over the wing (figure 3.3.2(b)) so that the vein 'BA' is parallel to the rungs of the ladder of the fan. Then the position of vein joint 'C' is noted and the measurement at the base of the fan is read to the nearest one tenth. The average for a bee sample (25-30 bees) collected from one hive is calculated and the range (highest and lowest) is included.

2) Discoidal shift: The discoidal shift is measured using a piece of white card marked with a 'T' shape. There are four gradations on either side of the vertical part of the 'T' at 2° intervals. The forewing is projected onto a wall and the horizontal line of the T is moved into position so that the line passes through the longest section of the radial cell with the vertical section passing through the joint marked 'H' (figure 3.3.2(c)). The position of the discoidal joint marked 'D' is recorded. If the discoidal joint is on the left side of the vertical line nearest to the wing attachment it is classed as negative. If the discoidal joint lies on the vertical line it is classed as zero and if the discoidal joint is on the right side nearest the tip of the wing it is classed as positive. Each sample is measured to the nearest degree for the sample of bees from the colony.

Once the cubital index and discoidal shift have been measured for a sample of bees from a colony then the results should be plotted on a scattergram with discoidal shift on the horizontal (x) axis and cubital index on the vertical (y) axis. The clustering of the data should provide an indication of the degree of racial purity of the sample.

3) Abdominal overhairs: Measures of overhairs are less important than the wing venation. The length of these overhairs is determined on the fifth upper abdominal segment (tergite). The length of these hairs can be compared to the width of a piece of 0.4mm thick wire. Short overhairs are up to 0.35mm long, medium hairs are 0.4mm and long hairs are more than 0.4mm. These lengths can be determined under a binocular microscope at 10x magnification.

4) Abdominal tomenta: The width of the bands of hairs (tomenta) that cross the abdominal segments on the third, fourth and fifth segments are measured. The fourth

Figure 3.3.2(a) Cubital index is determined by the ratio of the length of two veins A and B on the worker bees forewing, ratio A:B=cubital index (Redrawn from Dietz, A. 1993. *In*: The Hive and the Honey Bee. Dadant & Sons, Inc.).

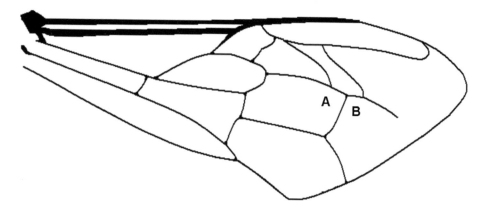

Figure 3.3.2(b) Position of the Herold fan for measuring Cubital index, the points 'A' and 'B' must lie on the vertical lines and then the position 'C' is recorded and read off a scale at the bottom of the Herold fan. Cubital index is the ratio of CB:BA (Redrawn from Dews, J.E. and Milner, E. 2004. Breeding better bees: using simple modern methods. Bee Improvement and Bee Breeders' Association)

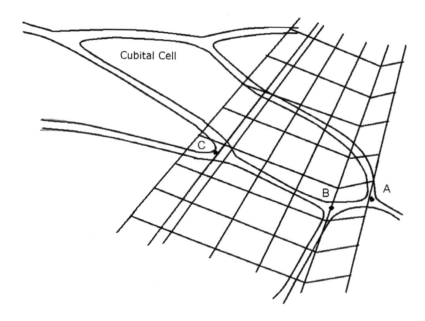

Figure 3.3.2(c) Discoidal shift is determined using a 'T' drawn onto white card. The forewing of a worker bee is projected onto a screen, the 'T' is placed over the radial cell so that the horizontal line joins the longest two points on the radial cell of the worker bees forewing, then the vertical part of the 'T' must bisect the connecting vein at 'H' at the middle of the radial cell, the position of point 'D' to the left or right of the vertical line determines the discoidal shift. In the diagram the discoidal shift is negative as the discoidal joint 'D' lies to the left (nearest the wing attachment) of the vertical line **(Redrawn from Dews, J.E. and Milner, E. 2004. Breeding better bees: using simple modern methods. Bee Improvement and Bee Breeders' Association).**

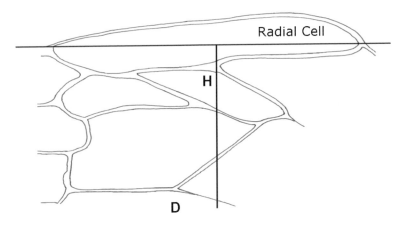

tergite is measured at the widest part. The tomenta may be grey or yellow (Italians). If the tomentum is less than half the width of the tergite the classification is narrow. If the tomentum is half the width of the tergite it is medium and if more than half the width of the tergite it is broad.

5) Abdominal body colour: If the colour of the second abdominal tergite is dark with spots less than 1mm², the bee is class B (black). If the bee has small spots larger than 1mm², the bee is class S (spots). If the spots have merged to become a band or ring the bee is class R (ring), with one band (1R), two bands (2R) or three bands (3R). Italian bees or hybrids have two or three yellow rings.

6) Length of the proboscis (tongue): The proboscis (prementum and glossa) is removed from the bee using jeweller's forceps, flattened on the stage of a binocular microscope and viewed through a calibrated eyepiece at 14x magnification. The order for the four races from longest to shortest tongue is Caucasian (7mm) > Carniolan (6.6mm) > Italian (6.5mm) > Dark European (6.0mm).

3.3.3 PROBLEMS WITH BEE BREEDING

Like all populations of animals, bees show great variation in colour, temperament, disease resistance and productivity. A successful beekeeper is always looking out for super-colonies that out-produce all other colonies.

Recording performance over two seasons enables hives to be evaluated as a source of potential breeders. But even with the great variety of stock available for breeding, little real progress has been made.

This is unfortunate as many of the basic operating costs involved in beekeeping are relatively constant regardless of the quality of bees that are used. A more productive bee can rapidly increase profits while operating costs remain constant.

Considerable effort has been made by beekeepers to breed bees with desirable characteristics. Many of these attempts have been unsuccessful largely due to the mating habits of the queen. For any breeding programme, no matter how sophisticated, the effectiveness will be limited by the following:

1. The queen may mate with as many as 12-17 drones and each drone may have a different genetic background
2. Queen bees have been known to mate with drones from hives up to 16 km away

3. A natural colony of bees cannot be looked upon as an individual or a family but rather a group of bees with the same mother and different fathers.

A colony of bees is a collection of a:
- mother (queen)
- 12-17 fathers (drones, now dead)
- daughters (workers)
- sons (drones)

Most of the workers in the colony are half-sisters, having the same mother, but different fathers. So the bee colony is regarded as a super-family in which a number of sub-families are mixed together. The queen is the mother of all sub-families, but each drone with which she mates is the father of only one sub-family.

Another difficulty relates to the breeding characteristics that are selected. Many of these characteristics involve worker behaviours, and many genes may be involved in these behaviours. Attempting to select just a few characteristics may not always be easy as often undesirable or unexpected characteristics may also be inherited.

Eliminating environmental influences, such as drifting, while sometimes very difficult is also important, otherwise selection may be based on characteristics that are not genetic and not inherited but influenced only by external environmental factors.

These limitations make breeding bees far more difficult and challenging than the breeding of livestock that can be confined and dealt with as individuals.

3.3.4 PREVENTING INBREEDING

An ideal breeding programme uses a large number of hives (100 - 400) on which to test the progeny produced from selected breeding stock. If only one queen mother hive is used as a breeder and the virgin queen bees produced from this hive mate with drones from the same hive, the inbreeding depression created will result in reduced variation in the offspring. Such an effect shows up as a patchy brood pattern laid by the queen. The patchy brood pattern is due to the loss of sex alleles as a result of the sex genes being alike. The larva is referred to as homozygous and the sex of the bee is male. The queen lays diploid male eggs in worker cells and when these eggs hatch the larvae are eaten by the workers, resulting in a spotted brood pattern (section 3.1.3).

In order to retain the sex alleles in a breeding population and prevent inbreeding, virgin queen bees should be reared from five to ten selected queen mother hives and mated in an isolated area to drones from ten to fifteen selected drone mother hives. The mated queens produced from isolated mating can then be used to requeen production hives and evaluated over the coming season.

A different set of queen mother hives and drone mother hives should be selected the following season. The same queen and drone mother hives should never be used from one generation to the next to minimise the loss of sex alleles.

In a 'closed' population breeding programme no drones are introduced from outside the breeding hives in the programme. This is maintained by isolated mating or instrumental insemination. Colonies may be bred for about 20 generations before the loss of sex alleles will cause inbreeding depression and up to 50% of the eggs will be diploid males. Eventually the colonies will collapse due to the lack of worker bees produced to sustain the colony.

3.3.5 ESTABLISHING A BREEDING PROGRAMME

A breeding programme requires:

i)   a queen rearing apiary where queens and drones can be reared
ii)   a home apiary for evaluating selected hives from the previous season on a monthly basis (figure 3.3.5)
iii)   isolated mating areas where feral bees are completely absent, where selected virgin queens in nucleus hives can be taken to mate with selected drone mother hives of a desirable breeding stock
iv)   a number of out apiaries for evaluating mated queens throughout the season
v)   accurate record keeping
vi)   an insemination laboratory (an alternative to an isolated mating area)

The home evaluation apiary contains colonies that have been selected from the previous season and, following further evaluation, some of these hives will become the queen and drone mother hives used for breeding during the next season.

The queen rearing apiary has cell-raising hives available for rearing new queens from grafted larvae. These hives need to be strong hives with abundant nurse bees that are well fed with pollen or pollen substitute, and honey or sugar syrup. At least one cell-raising hive should be available for every 20-30 queen cells that are reared. Cell

Figure 3.3.5 Home apiary for evaluating breeder queens using an individual queen record.

raising hives should be checked to ensure they feed copious royal jelly to young larvae. The queens that are reared for testing and evaluation should be the best queens that it is possible to rear under favourable conditions from the queen mother hives selected.

It is also important to provide abundant drones from selected drone mother hives to mate with virgin queens. Drone mother hives should be given abundant feed just as the queen mother hives are, and at least ten to fifteen hives should be selected. Drone comb should be placed in the middle of the brood, so that the queen is encouraged to lay on this comb. By careful attention to drone mother hives drones with plentiful semen will be produced.

At least 50 sexually mature drones should be provided in an open isolated mating situation to ensure adequate mating of each selected virgin queen bee.

3.3.6 SYNCHRONISING QUEEN AND DRONE MATURITY

Synchronisation of sexually mature drones and virgin queens is important for successful mating or insemination. There is a 21-day period from the time the egg is laid until sexual maturity for the virgin queen and a 36-day period from egg laying to sexual maturity for the drone. Therefore drone eggs need to be laid 15 days prior to queen eggs in order for the queens and drones to be sexually mature at the same time. Alternatively larvae less than 24 hours old (on the fourth day of development), should be grafted 19 days after the drone eggs have been laid in drone comb. Then the larva has 12 days till emergence as a queen (16 days total) and 5 days to become sexually mature, a total of 17 days (19+17=36 days). Both the virgin queens and drones would need to be confined to their hives by queen excluders on the entrance, or on the bottom board for drone mother hives. Banking of drones from hives of desirable stock may need to be considered.

Approximately four to five strong double hives, each with a frame of drone comb introduced, would be required to produce sufficient sexually mature drones to inseminate 50 virgin queen bees.

3.3.7 ESTABLISHING A MATING YARD

Before the grafted queen cells are ready to emerge, nucleus hives need to be established to accept the queen cells. Depending on the number of queen cells to go out, the nucs may be established on the tenth day after grafting, or the previous day. The nucs should have adequate frames of emerging brood, worker bees covering each frame, pollen, honey, suitable ventilation, and some form of queen excluder on the entrance to control the timing of mating.

Nucleus hives can be decorated with colours or symbols near the entrances to assist the queen bee to orientate after mating flights. Nucleus hives can be positioned in pairs with entrances facing in opposite directions. The nucs are spaced at least one metre apart and placed in an irregular pattern (see figure 2.5.10).

Natural landmarks can be used to assist the queen to orientate. Nucs should be protected from direct sun, ants, and strong winds. Record the date the cell is introduced on the lid of the nuc. Check the nucs three days after cell introduction to ensure the queen has emerged.

When bringing hives into an isolated area for mating, ensure the queen excluders are secured over the entrances and the lids make a firm seal. Ensure enough drone mother colonies are introduced for mating with virgin queens. The drone mother hives should be manipulated so that the maturity of drones and virgin queen bees coincides. When the drone hives and the nucs have been relocated to an isolated mating site and the virgin queens are five days old, the queen excluders on all hives can be removed on a warm, sunny afternoon with little wind. Check the nucs again after three weeks to determine if mating has occurred.

If a queen is mated and removed, draw a line through the date. If the nuc is queenless the date is circled. If the queen is in the nuc but not laying, the date is left unaltered.

3.3.8 ISOLATED MATING

Many attempts have been made to control the mating of virgin queen bees and drones. Attempts at mating in confinement, by the construction of large cages and releasing drones and virgin queens inside, have not been successful. The queen and drones appear to become disorientated in the confined space. Other attempts have included tethering queen bees to a mechanical arm then 'flying' the queen in circles in a cage full of drones. This has given mixed results.

Using the isolation offered by islands, mountainous regions or deserts where feral bees are absent, provides a greater chance of success. Desert areas, where bees do not normally survive, can be ideal for a one-off mating over a period of only a few weeks. In Australia Joe Horner, from New South Wales, regularly drove 1500 km in a round trip to the Hay Plains to obtain pure matings of different lines and races of bees.

Islands are also used in Australia. The Western Australian Queen Bee Breeding programme used Rottnest Island, off Fremantle near Perth, for isolated mating of Italian queens. On Kangaroo Island, near Adelaide in South Australia, there is a sanctuary for Ligurian (Italian) bees introduced in 1885 and maintained by the Ligurian Bee Act 1885 as a pure disease-free line. Islands off the coast of Queensland, including Bribie Island near Brisbane, have been used for mating Italian and Caucasian queens. In Canada mountainous areas have been used for isolated mating although these areas often have little forage for bees so feeding is required.

In any isolated area to be used for mating, an attempt to eradicate all feral colonies should be made before hives used for mating are moved into the area.

An alternative to isolated mating is to flood an area with drones of the desired stock so that the probability of a queen mating with the drones from hives, reared and selected by the breeder, is increased. Horner, in Australia, had some success in obtaining pure matings in a warm climate by flooding the area with drones released on a warm evening followed by virgin queens released about half to one hour before dusk. At this time of day many of the feral drones had stopped flying.

The final method of controlled mating is to instrumentally inseminate queen bees to guarantee the parentage (section 3.4).

3.3.9 SELECTIVE BREEDING PROGRAMMES

Successful selective breeding programmes have four essential components:

1   Stock selection: In order to select a colony there must be selectable differences between the superior colony and other colonies resulting from the parent population.

2   Genetic variability: The differences between colonies must be due to genetic differences of the parents e.g. queens and drones. This is why it is important to eliminate such influences as drifting which may cause differences between colonies that are not genetic - they are related to the environment and are not inherited. It is important to control all environmental variables as much as possible so that variability is due to genetic differences only.

3   Controlled mating: When superior colonies are identified the progeny from these superior colonies must be mated under controlled conditions to avoid unwanted genetic material entering the breeding programme as occurs when males from hives not selected mate with queens inside the programme.

4   Stock maintenance: Superior stock must be maintained from one season to the next otherwise the selective breeding efforts will be lost.

There are two main types of breeding programmes:

1   **Inbred hybrid breeding:** Different inbred lines of bees are bred. These lines display a loss of vigour apparent from the speckled brood pattern due to the reduced sex alleles in the lines. When an inbred line is crossed to another inbred line the resultant hybrid vigour is displayed by superior characteristics. The individual lines are bred initially from superior stock and are selected on their combining ability with other lines. The difficulty with this programme is that maintaining inbred lines may be difficult once inbreeding depression is significant i.e. egg viability is 50% or less. This occurs when only two sex alleles remain in an inbred line.

2   **Closed population breeding:** The objective of a closed breeding programme is to progressively improve a population of bees by selection while maintaining high brood viability and genetic variability. The population remains closed to mating from bees outside the population. Matings are controlled by allowing the queen to only mate with drones in an isolated area free of drones from other hives outside of the closed population, or by controlling mating using instrumental insemination e.g. Western Australian Queen Bee Breeding Programme. Breeding programmes in a closed population may include:

a)   Identifying superior colonies and selecting 35-50 breeder colonies to be maintained each generation. Daughter queens are raised from all breeder hives and mated with drones from the same breeder hives using either instrumental insemination with drones selected at random or using an isolated mating area.   Mated daughters are placed in hives for evaluation over the following season and at the end of the season the top 35-50 colonies are selected again. This ensures brood viability of 85% or more is maintained for 20 generations.

b)   Similar to the above except select 25 breeder queens from different apiaries.   Each breeder produces daughters and these daughters replace all hives in the apiary from which they came. Fewer colonies are required with only 25 breeders, and brood viability of 85% or more is maintained.

c)   Semen from drones is collected from each breeder queen as above. The semen is homogenised and used to inseminate all the daughter queens bred from each breeder queen. The inseminated queens are obtaining genetic material from the entire gene pool within the closed population. The off-spring should have less variability in brood viability and there should be more sex alleles maintained in the population.

3.3.10 VARROA TOLERANT BREEDING PROGRAMME

There are a number of approaches to breeding colonies that are tolerant to varroa mite. Three methods are discussed.

The first method is to:

Select good honey producing hives that also show (a) hygienic behaviour with workers able to uncap and remove dead brood and (b) high levels (up to 40% or more) of suppressed mite reproduction (SMR) in infested brood and (c) low levels of mites per 100 worker bees. Before the autumn treatment these hives should have good brood, few worker cells with mite faeces, few drone cells infected with mites, few deformed wings in adults, and few dead mites on the bottom board.

The second method is to:

Select hives that are good honey producers that have been left in an apiary untreated for 12 months either by accident or on purpose, as well as locating abandoned or untreated apiary sites and selecting hives that are still alive and have produced surplus honey. From 3-10% of these hives may show varroa tolerance.

The third method is to:

Select good honey producing hives from the previous season and treat all these hives with a varroa treatment in spring. Allow these hives to gather honey throughout the season with no further treatment. At the end of the season when there is little brood in the hives, determine the mite fall from all hives over a 7-10 day period using sticky boards. Score the hives according to mite fall and select those with the lowest mite fall.

Using one of the above methods of selection, approximately 10 hives should be identified and placed in an isolated apiary, 5km away from other managed hives that are being treated for varroa. These 10 hives are then monitored every 3 months for varroa levels by sampling 100 workers, washing these workers in soapy water or alcohol, and counting the number of mites per 100 bees. Colonies with more than 15 mites per 100 bees should be either removed from the isolated apiary or requeened with queens produced from hives with less than 15 mites per 100 bees. This level can be reduced to 10 mites per 100 worker bees over several seasons.

Poor honey producing hives and aggressive hives should be removed from the isolated apiary. All queens should be marked and drone comb added to each of the hives meeting the selection criteria. Queens can be raised from the most tolerant hives and the cells placed into nucs that are located near the isolated apiary so that the virgin queens mate with drones that carry varroa tolerance genes.

The mated queens produced can then be used to requeen production hives and then evaluated again at the end of the season before autumn treatment. Selected hives are then returned to the isolated site where further selection and queen raising continue. Feral colonies in the isolated area should not be of concern as they will have only varroa tolerant drones surviving.

If an isolated site is not available then queens will need to be instrumentally inseminated with drones selected from varroa tolerant hives.

3.4 Instrumental Insemination

3.4.0 INTRODUCTION

One of the difficulties in breeding bees with characteristics desirable for the beekeeper's requirements is the problem associated with controlling mating. It is not possible to place a fence around the bees and prevent the drones and queens from leaving a given area as is the case with sheep and other domestic animals.

The drones and queens may fly over large distances, with queen bees recorded mating with drones from hives separated by up to 16 km. Also the queen may mate with 12-17 drones from different hives and of completely different genetic stock (section 3.3.3). Effectively this means that there is little control over what the offspring may be like.

For more than a century beekeepers and scientists sought a way to control the mating of their breeding stock. Efforts to control parentage have been made by controlling natural mating and by working out methods of artificial insemination. The latter approach was developed in two ways by:

i) attempting to introduce semen directly from the copulating organs of drones into the queen sting chamber and

ii) by utilising instruments to inject the semen into the oviduct of the queen. This method is called instrumental or artificial insemination.

3.4.1 BENEFITS OF INSTRUMENTAL INSEMINATION

Instrumental insemination has a number of **advantages** over open mating of queens. The mating of the queen mother's daughters can be controlled so that evaluation of the progeny and subsequent selection is more meaningful. Specific matings can be made between certain breeder hives with desirable characteristics and the offspring evaluated. Some crosses may be desirable while others may have less suitable characteristics. Virgin queen bees may be inseminated day or night in good or bad weather. Each queen inseminated should have a normal or near normal number of spermatozoa in the spermatheca and theoretically should be able to outlast the productive life of the field-mated queens.

There are a number of **disadvantages** of instrumental insemination.
Drone supply: There is a need for a continuing, abundant supply of healthy, sexually mature drones during the entire insemination period.

Skill Levels: In order to perform insemination of queen bees there is a requirement to train and practice the techniques as it is precision work which some people may find difficult to undertake over an extended period. Hence having skilled operators can be a problem.

Sanitation Levels: Because of the invasive nature of the insemination process there is a need for a high level of sanitation and to work in a sterile environment.

Equipment: Appropriate facilities need to be set aside for the purpose of insemination.   The cost of the equipment required may make the insemination technique prohibitive for many smaller beekeeping operations.

3.4.2 INSEMINATION EQUIPMENT

A number of different insemination instruments exist on the market. The Mackensen is one of the older models from which other designs have been developed. The Schley instrument has ball and socket joints, a micro-manipulated syringe for fine adjustment, detachable capillary tubes, and tips with latex connectors that can be useful for transporting semen. This instrument can be connected to a Harbo large capacity syringe with a micrometer for accurate semen measurement. Other instruments including the Swienty and Vesley are not so popular but there is a move towards simple insemination instruments including the Kuhnert-Laidlaw, Jordan-Pollard and Latshaw devices which all require the use of forceps instead of the sting hook.

To achieve a sterile environment it is preferable to have a room with washable walls, and a floor and ceiling of sufficient size to accommodate the inseminators and the drone flight cage. The room or laboratory needs to have adequate bench space, good lighting and preferably have hot and cold water. Relative humidity should be maintained above 50% to prevent the queen's organs from dehydrating.

The following is a checklist for the equipment required:

- Dissecting binocular, stereoscopic microscope 10-40x magnification (1-2x eye objective, 10-20x stage objective) (figure 3.4.2(k))
- Cold light source (figure 3.4.2(l))
- Drone flight cage (50 cm long x 40 cm wide x 40 cm high; a bee box is ideal) with a heating element on the floor and a light source above e.g. desk top lamp (figure 3.4.2(r))

Figure 3.4.2(a) – (f) Instrumental insemination apparatus.

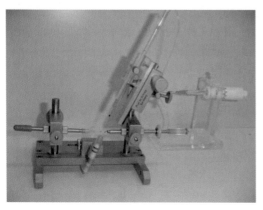

(a) – Schley insemination apparatus: stand with hook holders; from left ventral hook, queen holder, pressure grip forceps, syringe block with Harbo syringe containing glass barrel and plastic tip (above), micrometer (far right)

(b) - Hook holder connected to the stand at the base; and end of ventral hook passing through ball joint

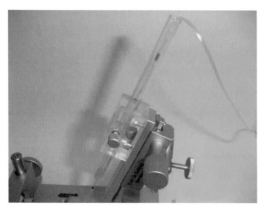

(c) – Syringe block: Harbo syringe with Tygon tubing leading to a protective glass barrel containing capillary tube and plastic syringe tip at lower end

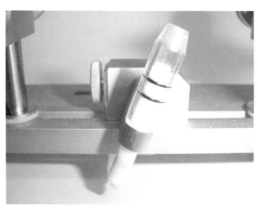

(d) - Queen block with plastic queen holder and rubber stoppers

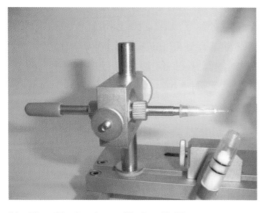

(e) – Ventral hook and queen block and holder

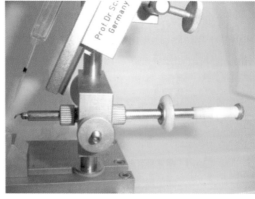

(f) – Pressure grip forceps and plastic syringe tip above

Figure 3.4.2(g) – (l) Instrumental insemination apparatus.

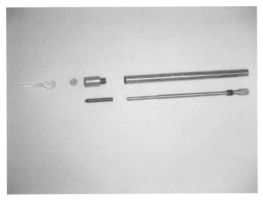

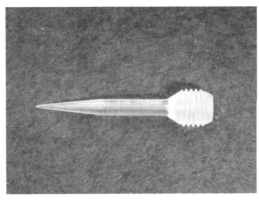

(g) - Mackensen syringe: (left to right) Plastic syringe tip, rubber diaphragm, metal adapter, short plunger (below), barrel, long or screw plunger (below)

(h) – Plastic syringe tip (each gradation is one microlitre)

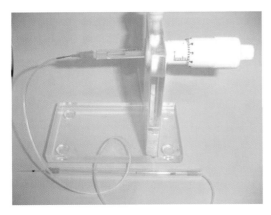

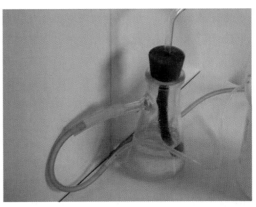

(i) – Micrometer for delivering 8 microlitres of semen into queen, with Tygon tubing and protective glass barrel containing capillary tube and plastic syringe tip

(j) – Wash bottle filled with water, used for controlling the CO_2 gas flow rate to the queen

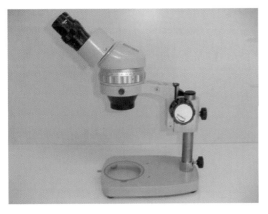

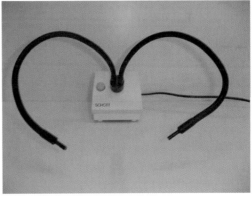

(k) – Binocular stereoscopic microscope

(l) – Cold light source with flexible arms that can be bent into position

Figure 3.4.2(m) – (r) Instrumental insemination apparatus.

(m) - Carbon dioxide bottle with regulator and coarse adjustment (right hand black knob) and fine adjustment (left hand small black knob)

(n) – Insemination apparatus and microscope set up

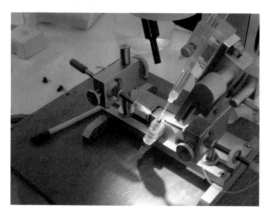

(o) – Inseminating queen: ventral hook on left side and pressure grip forceps on right side with Harbo syringe and plastic syringe tip used for insemination

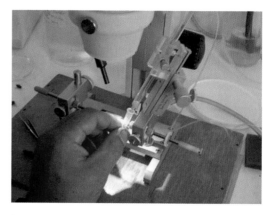

(p) – Taking up previously collected drone semen into syringe tip and capillary tube

(q) – Autoclave for sterilizing instruments

(r) –Drone flight cage and drone catcher box

- Drone catcher boxes (15 cm x 20 cm x 10 cm high) for collecting drones from hives (figure 3.4.2(r))
- Insemination apparatus (figure 3.4.2(a -f))
    - stand and hook holder (b)
    - ventral hook and hook holder (e) (sting hook optional)
    - syringe block (c)
    - queen block and holder (d)
    - pressure grip forceps (optional) (f)
    - Insemination syringe –
        - Harbo syringe (figure 3.4.2(c, f, i)
            - micrometer used for taking up and delivering known amounts of semen (i)
            - Tygon tubing leading to the capillary tube housed in a protective glass barrel with latex connecting tubing (c) and (i)
            - plastic syringe tip (f)
        - Mackensen syringe (figure 3.4.2(g-h))
            - plastic syringe tip (h)
            - rubber diaphragm (g)
            - metal adapter (g)
            - short plunger (g)
            - barrel (g)
            - long or screw plunger (g)
- Jeweller's forceps
- Carbon dioxide cylinder with regulator (figure 3.4.2(m))
- Rubber tubing
- Wash bottle - flask with rubber bung and glass inlet and outlet connected to rubber tubing (figure 3.4.2(j))
- Autoclave or frying pan with water for instrument sterilisation (figure 3.4.2(q))
- Physiological solution containing antibiotic e.g. gentamycin (10ml) and buffer solution e.g. Tris buffer (1 litre)
- Broad spectrum virucidal disinfectant e.g. Virkon
- Sodium hypochlorite solution e.g. Janola
- Beakers - 50ml
- Pipettes for washing syringe tips and rubber ends
- Queen mailing cages
- Paper towels, tissues, cotton buds, sterile wipes, cleaning rags, mops, etc

Before insemination begins, the walls, floor, ceiling and bench top should be washed with a virucidal disinfectant to sterilise the room.

The microscope should be set up at a comfortable height for viewing when sitting at the bench. The insemination apparatus is placed either directly on the bench top or on the dissecting stage of the microscope. The stand should be solid so that it does not move.

All metal that is likely to come into contact with the queen should be autoclaved including the ventral and sting hooks, forceps, metal plunger and metal adapter.

The plastic syringe tip is washed in neat sodium hypochlorite by placing in a beaker of solution. The syringe is then assembled. In the case of the Mackensen syringe the long plunger is inserted in the barrel, the short plunger is inserted from the opposite end, and the metal adapter is screwed into position.

The antibiotic is now mixed with the Tris buffer. This is usually at a ratio of 1:100 i.e. 1 ml antibiotic to 100 ml of Tris buffer solution. The resulting physiological solution is then taken up in a Pasteur pipette and the plastic tip of the syringe is washed and back-washed with solution. Using sterile forceps the rubber diaphragm is washed in solution and then fitted into the metal adapter. The adapter is then flooded with physiological solution and the plastic tip is then screwed into the adapter until it sits firmly against the rubber diaphragm.

The plastic calibrated tip of the Mackenson syringe (figure 3.4.2(h)) should now be fill with physiological solution and part of this must be ejected before semen can be drawn up into the tip. The fluid is moved down the tip until it reaches the last calibration mark. The solution should be withdrawn up to the ten microlitre (μl, 10^{-6} litres) mark. If this is not possible then further solution should be ejected from the tip and wiped off with a cotton bud moistened with physiological solution. Once the tip is ready the Mackensen syringe (figure 3.4.2(g)) is then mounted in the open jaw of the syringe block and the adjustment screw is tightened. The microscope is now focused on the tip in preparation to take up semen and the cold light source is directed towards the tip.

If the Harbo syringe is used, the plastic syringe tip is connected to the capillary tube using latex connecting tubing. The capillary tube is housed inside the protective glass barrel and connected (via a latex connector) to Tygon tubing which is in turn connected to the micrometer (figure 3.4.2(i)). The syringe, capillary tube and Tygon tubing is flushed with physiological solution and the syringe tip sterilised with sterile wipes. The glass barrel is then mounted in the syringe block (figure 3.4.2(c)) of the Schley instrument and is ready to take up drone semen (figure 3.4.2(p)).

3.4.3 QUEEN ANATOMY AND INSEMINATION PROCEDURE

The anatomy of the queen has presented problems for inseminators in the past. The reason for this difficulty is a flap, known as the valve fold (see figure 1.3(a)) at the base of the vaginal passage, which blocks the movement of semen. When semen is injected into the vaginal orifice the valve fold prevents this semen passing into the median oviduct and finally into the spermatheca, the sperm storage sac, where it is stored until required. The spermatheca is barely the size of a pinhead but stores 5.3 to 5.7 million spermatozoa.

Inseminators in the past have found that when injecting semen from the syringe, the semen may back up and not enter the median oviduct. This was overcome by the use of two hooks, the ventral hook and the sting hook, to open the sting chamber sufficiently for the syringe, when inserted, to by-pass the valve fold. More recently jeweller's forceps have been used to replace the sting hook. The inseminator pulls the sting in a dorsal direction while manipulating the syringe to by-pass the valve-fold. Most recently, pressure grip forceps have been used (figure 3.4.2 (f)). These forceps are mounted in the sting hook holder and by applying a slight pressure the forceps are opened and the sting of the queen can be clasped and then pulled in a dorsal direction to open the sting chamber. Once the sting chamber is open there is not the requirement to keep holding the pressure grip forceps, as is the case with the jeweller's forceps, the latter method needing a very steady hand.

3.4.4 DRONE ANATOMY AND SEMEN COLLECTION

The testes of the drone supply semen via the seminal vesicles and ejaculatory duct to the bulb of the penis or endophallus (see figure 1.3(b)). Drones of 12 days old or more are sexually mature and may provide semen when their endophallus is everted. Drones beyond 20 days old are unsuitable.

To collect sexually mature drones a drone mother hive, with queen excluders on the bottom between the base and bottom brood box and on the top of the hive under the lid, should be opened in the morning before the temperature increases above 15°C. Mature drones are usually found on the outside two frames in the brood nest and at the entrance to the hive. Remove as many drones as necessary i.e. 30-50 per queen to be inseminated and place in catcher boxes (figure 3.4.2(r)). Transfer catcher boxes to the insemination room. Transfer drones into a flight cage provided with a heating element on the floor and a warm light source above to maintain the temperature at 20°C or more to keep the drones actively flying.

Alternatively a catcher box with a queen excluder lid and a sliding panel on the bottom is placed on top of a drone mother hive after the lid is removed. Drones attracted to the light will fly up into the catcher box and workers will pass through the excluder. The sliding panel is then slid back across catching the drones inside.

The flight cage (a bee box with an excluder lid and a wooden base is ideal) should have a hole, large enough to place a hand through, on one side. The hole is covered with overlapping rubber sheeting. Drones cannot escape through the rubber sheeting but the inseminator can reach in and pluck the drones off the queen excluder. Active drones ejaculate more readily than sluggish drones. A final selection of drones that are active and the appropriate colour, can be made at this stage.

Before obtaining semen, your hands should be washed thoroughly. Remove a drone from the flight cage and hold the drone upside down with the head between thumb and forefinger. At this point some drones may evert the penis and ejaculate semen. However, most drones require further stimulation. The thumb should now squeeze and crush the head and thorax in an effort to stimulate partial eversion, thus sacrificing the drone. If this is unsuccessful apply weak pressure on the abdomen laterally (on the side) this should result in the yellow bursal cornua of the drone appearing. For full eversion roll the fingers along the abdomen starting from the anterior base and working towards the tip. This should result in full eversion, with a ball of white mucous appearing on the bulb of the endophallus (see figure 1.9(b)). Semen should appear on the tip of the mucous layer as a slightly pink to orange colour and forms a lattice work in comparison to the very white ball of mucous.

It is important to ensure the mucous and semen does not touch your fingers or the abdomen of the bee as contamination will occur.

To collect semen hold the bulb of mucous up to the tip of the plastic syringe. By moving the tip of the syringe down just above the semen layer touch the syringe tip on the semen and start to withdraw semen by winding the end of the long plunger so as to take up semen. At the same time roll the drone over slightly in order that all the semen is removed. Care should be taken not to take up mucous as this will block the tip. Maintain a small air space between the semen and the physiological solution. Each drone will produce about 1microlitre (μl) of semen.

Continue to evert further drones and collect semen by first ejecting a small amount of semen onto the semen of the next drone and then taking up the remaining semen. About 8 microlitres of semen is required for each insemination whereas the tips will take up to 10 microlitres.

After 8 μl has been taken up, withdraw the semen a millimetre further to prevent the formation of a seal at the end of the syringe tip. The tip should now be dipped in physiological solution and about 0.5 microlitres withdrawn. This serves to prevent infection, to prevent the formation of a seal at the end of the syringe tip and to act as a lubricant when the tip is inserted into the queen. The tip should now be wiped with a cotton bud dipped in physiological solution; discard the cotton bud. The semen is now ready for insemination into the queen.

3.4.5 MOUNTING QUEEN IN HOLDER

Queen cells are placed into nucs that have queen excluders covering the entrance. Five days after emergence, the virgin queens are removed and placed into JZ's BZ's, Miller or mailing cages. The cage and the nuc are marked with the same number so that the queen can be returned to the same hive after insemination. The caged queen is then transferred to the insemination room. The cage is placed into a container and food grade carbon dioxide is pumped through a rubber hose into the bottom of the container until the queen has been anaesthetised.

Once the queen is still she can be removed from the cage by hand and carefully guided headfirst into a small temporary holding tube with a small hole at one end. With the queen inside the holding tube, the queen holder, a plastic length of tube 6.6 mm diameter tapering to 4.8 mm at one end, is held against the tube and the queen is blown backwards with a short puff into the queen holder so that her abdomen protrudes from the tapered end.

The queen holder is placed onto the queen block (figure 3.4.2(d)) and the outlet rubber pipe leading from the carbon dioxide (CO_2) regulator (figure 3.4.2(m)) is connected via a wash bottle (figure 3.4.2(j)) to the rear of the queen block. The flow rate of gas can now be regulated by observing the rate of bubbles that leave the tubing submerged in water inside the wash bottle. A rate of approximately one bubble per second is adequate. The CO_2 gas is supplied from a small high pressure cylinder supplied with a regulator with a pressure gauge. It is preferable to have two discharge valves or outlet pipes so that anaesthetising the queen, and inseminations, can be performed simultaneously. If the queen starts to move while mounted in the queen holder, the rate of CO_2 can be increased.

Figure 3.4.2(s) Sting chamber of the queen open for instrumental insemination (Redrawn from Dade, H.A. 1985. Anatomy and dissection of the honey bee. International Bee Research Association; and Ruttner, F. 1976. The instrumental insemination of the queen bee).

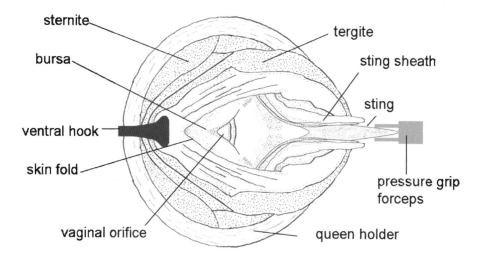

The queen holder is mounted on a 32° angle from the vertical to allow for ease of insemination, with the syringe mounted at the same or a slightly greater angle. The queen is mounted with the ventral surface (lower side) facing towards the vertical or towards the left ventral hook.

The ventral hook is lowered into the sting chamber against the ventral plate until the tip touches the anterior membrane (figure 3.4.2(s)).

Pressure grip forceps are mounted in the sting hook holder (replacing the sting hook) and manipulated into position to firmly grasp the sting that is then pulled in a dorsal direction (to the right) to open the sting chamber. With the sting chamber open the valve fold is withdrawn sufficiently to allow semen to be injected into the median and lateral oviducts. Alternatively jeweller's forceps are used to grasp the sting and open the sting chamber. Both methods replace the use of the sting hook.

The vaginal orifice and vaginal passage is located, as well as a 'V' formed on the ventral side of the vaginal orifice. The syringe is manipulated dorsally into this 'V' and the tip inserted so as to by-pass the valve fold using a zigzag action. Once the syringe tip is in position (figure 3.4.2(n) and (o)) inject a little physiological solution. If the solution appears to enter uninhibited continue to inject the remaining semen. If the solution does not enter easily, reverse the tip back slightly and repeat the process. Eventually all the semen should enter the vaginal orifice without backing up.

If semen continues to back up, the syringe tip should be withdrawn and the tip and any semen in the sting chamber should be wiped clean with a sterile wipe or a cotton bud or tissue paper soaked in physiological solution. The syringe should be filled with more semen (figure 3.4.2(p) from further drones, or previously collected, until the syringe has the required 8 microlitres for insemination. The syringe should then be reinserted and a further attempt to inject semen repeated as before.

Over the next two days the sperm injected into the queen is forced back out of the lateral and median oviducts by the queen and into the spermatheca. The valve fold directs the semen into the spermathecal duct and eventually into the spermatheca. At best, after a migration period of 40 hours, only 10% of the sperm will reach the spermatheca, the rest being eliminated by the queen. The migration of semen from the oviducts into the spermatheca is influenced by the temperature surrounding the queen for the first few hours following injection of the semen; and the activity of the queen. This is why the queen should be kept warm after insemination and returned to the hive soon after she has recovered from the anaesthetic.

Once the queen is inseminated she can be removed from the queen block by lifting the queen holder off the stopper. By gently blowing the queen into cupped hands the wings of the queen are clipped (left for odd years, right for even) and marked by gluing a coloured disc, appropriate to the year she is inseminated, on her dorsal thorax. The queen is then placed back into a queen cage to recover. The queen should be kept warm (20°C) during this period. Once the queen is walking, place the cage back into the original nuc between two brood frames.

3.4.6 MONITORING INSEMINATED QUEEN BEES

After 24 hours the caged queen is removed again from the nuc, returned to the insemination room and the queen is gassed again with CO_2. The second anaesthetic initiates oviposition (egg laying) and can be undertaken 24 hours before or after insemination. The caged queen is then allowed to recover in a warm environment and when she is active the cage is returned to the nuc and the queen is released slowly by allowing the workers to chew through the candy over the next 24 hours.

The inseminated queen should be checked to determine if she has initiated egg laying three to five days after her second dose of CO_2. If insemination has been successful the queen should lay an even pattern of worker brood.

An insemination record, or log book, should be maintained to record the following information:

- date queen cell introduced into nucs
- dates of drone and virgin queen sexual maturity
- the parentage of both queen and drones if known
- the date of insemination
- the date when the second dose of CO_2 is applied
- the date when the queen begins to lay eggs and
- the hive number and the number and colour of the disc attached to the queen.

3.4.7 ENCOUNTERING PROBLEMS

If the insemination laboratory or equipment is not sterile the queen may become infected. In this case the queen may develop septicaemia (blood poisoning) and may fail to lay or may begin laying then die soon after. If this occurs check the disinfection and sterilisation procedure of the laboratory and equipment as well as the age of the antibiotic and the method of semen collection.

Inseminated queens should be left in nucleus hives until egg laying is well underway before the nuc with the new queen is united into a full size hive. Otherwise the queen may be superseded. The nucleus hives should be fed sugar syrup (1:1 sugar: water) and pollen substitute to encourage queen egg laying.

After the inseminated queens have laid, and been united with newspaper into full size hives, the queen's progeny and performance can be evaluated just as for potential breeder queens.

3.4.8 SEMEN MIXING AND STORAGE

Some inseminators have used the technique of homogenising semen from drones collected from the best performing hives in a breeding programme. The benefits of this technique include: (i) increasing the gene pool and maintaining variation from all the selected hives in a closed population; (ii) reducing variability in brood viability; (iii) maintaining sex alleles in the population; (iv) and because all queens are inseminated with the same homogenised semen, variation in performance is due mainly to the maternal side.

The semen is collected in small plastic vials and then transferred into capillary tubes (see figure 3.4.2(c) and (p)) that are first cleaned in physiological solution containing antibiotic. The capillary tubes are sealed at both ends with vaseline petroleum jelly close to the column of semen. The tubes are kept at room temperature (13-15°C) away from sunlight. These tubes can be transported anywhere in the world with little danger of introducing bee diseases. The semen will last for several weeks.

When all the semen has been collected it is introduced into small glass vials and then spun in a centrifuge at high speed to thoroughly mix the semen. The semen can then be drawn off into plastic syringes as required and up to 50 or more inseminations can be undertaken per day.

Acknowledgements

I wish to thank Dr Nelson Pomeroy, Jan Johnstone, David Yanke, Jan McAuslan and Lynne Johnston for providing critical feedback on the manuscript.

I wish to thank Angela and Jacqueline Woodward, and Jan McAuslan for assisting with editing of the manuscript.

I wish to thank Iain Winslade for assistance with the graphic design in the manuscript and Sharon Goatley for computing assistance.

I wish to thank the students who have studied queen bee rearing over the last eight years for their enthusiasm and suggestions on course improvements and for the following students that gave permission to have their photos appear in the manuscript: Grenville Watson, Cory Rusbatch, Andrew Velman, Sam Bell-Iyer, Jude Sharpin, and Daniel Adams.

I wish to thanks Frans Laas for the use of equipment for instrumental insemination.

I wish to thank Telford Rural Polytechnic for providing the opportunity to complete this book.

General Glossary

alleles	alternative forms of a gene
American foulbrood	bacterial disease of honey bee brood
body cell	involved with the structure and function of an animal
cell	wax constructed cavity located on the comb used for rearing bees; smaller hexagonal (horizontal) cells are used for rearing workers while drones are reared in larger hexagonal (horizontal) cells; both types are used for storing pollen and honey; queen cells are rounded, hang vertically with a tapering blunt tip
chalkbrood	fungal disease of honey bee brood
chemoreceptors	sensory organs used for detecting pheromones
chromatid	one of the two strands which result from duplication of the chromosome that become visible during mitosis and meiosis
chromosome	thread-shaped body, consisting usually of DNA
crossing over	exchange or recombination of genetic material in the chromosomes during meiosis
diploid	two sets of chromosomes, one derived from each parent
DNA	deoxyribonucleic acid; substance that chromosomes are made of
dominant	where one gene of a pair has influence over the other
drones	male honey bees that mate with the queen and then die
endomitosis	replication of one set of chromosomes into multiple copies without division of the nucleus, such as occurs in drones, where the number of chromosomes doubles
European foulbrood	bacterial disease of honey bee brood; not present in New Zealand
extrafloral nectaries	nectaries that occur outside the flower
fertilisation	joining together of gametes i.e. the spermatozoon (sperm) and the ovum (egg)
genes	units of inheritance; short length of chromosome, influencing a particular set of characters in a particular way
gametes	reproductive cells, sperm and egg
grafting	transfer of larvae from brood frame into cell cups using grafting tool
gustation	detecting pheromones in liquid state
haploid	one set of chromosomes, usually derived from one parent
heterozygous	genes of a pair containing different information

homologous	pair of chromosomes of the same size and shape, one member of the pair comes from the mother, the other member of the pair comes from the father
homozygous	genes of a pair containing the same information
honeydew	sugary excretion from sap sucking insects collected by bees and other insects and bought back to the hive and stored like honey
independent assortment	when chromosomes line up at the equator of the cell in a way that is totally random
locus (s) loci (pl)	position occupied by a particular gene on the chromosome
mating sign	remains of a drone's penis inserted into the back of the abdomen of the mated queen
meiosis	reproductive cell replicates producing gametes
mitosis	body cell replicates; simple cell division; the new cell is identical to the parent cell, no gametes are formed
nectary	area inside flower where nectar is secreted
olfaction	detecting pheromones in gaseous state
oviposition	egg laying
parthenogenesis	development of an egg without fertilisation into a new individual such as occurs when the queen lays unfertilised eggs that develop into drones
pheromone	chemical messages secreted externally resulting in a behavioural response in an animal of the same species
piping	series of high pitched sounds produced by the queen causing workers to freeze on comb
polarise	confining light travelling in many directions into travelling in one direction
propolis	resinous secretion from plants that is scraped off by workers and brought back to the hive to gum up cracks and sterilise cells
queen	there is usually one queen in a hive, laying fertilised and non-fertilised eggs that develop into female and male bees respectively; she also produces queen substance which affects the behaviour of worker bees in the hive
queen substance	pheromones produced by the queen that regulate the behaviour of workers
queenless	hive without a queen
queenright	hive with a queen

recessive	where one gene of a pair does not have influence unless occurring in a homozygous pair
reproduction	method by which genetic information is passed to the next generation
reproductive cell	involved only with reproduction; cells divide to produce gametes, sperm and ova
royal jelly	secretion from the hypopharyngeal and mandibular glands of workers that is fed to queen larvae
sacbrood	viral disease of honey bee brood
septicaemia	blood poisoning
workers	female non-reproductive bees that make up most of the hive bees, and who perform hive duties and forage for nectar and pollen
worker jelly	secretion from the hypopharyngeal and mandibular glands of workers that is fed to worker larvae
zygote	cell produced during fertilisation when the egg and sperm combine

(s)* = singular or one; (pl)** = plural or more than one

Glossary of Anatomy and Development

abdomen	largest hind region of the bee that contains the sting at the tip.
antenna (s)* antennae (pl)**	feeler(s)
antenna cleaner	notch in front legs for cleaning pollen off antennae
arolium	suction pad at tip of leg that allows bees to walk on smooth surface
auricle	pollen press on hind leg of worker bee
basitarsus	swollen part of tarsus on hind leg
bulb, cervix, lobe, bursal cornua	parts of penis in drones
callow	newly emerged adult
chorion	outer shell of egg
claw	at tip of leg used for holding onto rough surface
corbicula (s) corbiculae (pl)	pollen basket(s) made of hairs on outside of hind legs of workers
corpora allata	relatively large globular organ found on the sides of the oesophagus of both larvae and adults
coxa, trochanter, femur, tibia, tarsus	parts of the leg
crop	another name for honey stomach
cuticle	hard outer layer of insect, impermeable to water and air
dorsal	upper side of bee
ejaculatory duct	canal that mucous and sperm pass through on the way to the penis
embryo	cells developing inside egg after fertilisation to form new larva
exoskeleton	outer skeleton used for protection and muscle attachment
facets or ommatidia	small separate elements, each with light sensitive cells which collectively make up the compound eye of the bee
flabellum	tip of glossa part of proboscis
flagellum	segmented distal (tip) of antenna
galea	frontal pair of flaps making up the tongue
glossa	the hairy tongue that moves up and down the tube of the proboscis
haemolymph	blood of the bee
hamuli	row of hooks on leading edge of hind wing which couple with

	a curved fold on the trailing edge of forewing and enable the wings to beat together
hypopharyngeal glands	brood-food glands in head that produce royal jelly and enzymes in workers
instar	stage of larval development
inter-segmental	between the segments of the body, each segment is covered above and below by hard plates; soft membrane exists between the plates
juvenile hormone	JH; involved in caste determination in honey bees; and known to regulate growth, development and metamorphosis in many insects
labial palps	rear pair of flaps making up the proboscis
lateral oviduct	paired oviduct arising from the two ovaries of the queen
lancet	barbed part of sting
larva (s) larvae (pl)	the grub stage when feeding takes place
mandibles	jaws
mandibular glands	involved in softening wax and royal jelly production in workers and production of queen substance in the queen
median oviduct	common oviduct that leads from paired oviduct to vaginal chamber in queen
metamorphosis	change of form e.g. from larva to adult
micropyle	small hole in egg through which sperm enters
midgut	ventriculus, main digestive stomach where food is absorbed
mucous glands	glands that produce mucous for sperm to travel through in drone
Nasonov or scent gland	at tip of upper abdomen attracts other bees when released by workers
ocellus (s) ocelli (pl)	simple eye(s) on the top of head, probably to detect light intensity
oesophagus	part of gut system between mouth and stomach
ovaries	two large organs that contain ovarioles in queen and worker
ovariole	thin tubule inside ovary of queen or worker where egg is produced
oviduct	tube through which egg passes before reaching vaginal orifice
ovum (s) ova (pl)	another name for egg
prementum	first part of proboscis that connects to the back of the head
prepupa	stage between larva and pupa when larva stretches out

	along cell, the cell is capped and the larva spins a cocoon
pro, meso, meta thorax	three segments of thorax
proboscis	tongue
pupa (s) pupae (pl)	stage when larva changes (metamorphosis) from a larva to an adult
rastellum	a rake of hairs on worker; when the two back legs rub together the rake picks up pollen from the combs of hairs on the opposite leg
salivary glands	includes cephalic labial and thoracic labial glands involved in the processing and digestion of food
scape	first basal part of antenna
scopae	rows of hairs on inside basitarsal segment of hind legs of workers
seminal vesicle	enlarged sac that sperm passes into from testes in drone
silk gland	secretes silk in the larva through the spinneret
spermatheca	storage sac in queen for storing sperm
spermatozoon (s) spermatozoa (pl)	another name for sperm
spinneret	lobe formed at the mouth of the larva connected to the silk gland and used for spinning a silk cocoon prior to pupation
spiracles	breathing holes in side of exoskeleton in thorax and abdomen
sternites	lower (ventral) plates of abdomen
stylet	cover which forms canal for poison sac
tergite glands	along upper abdomen that secrete queen substance
tergites	upper (dorsal) plates of abdomen
testis (s) testes (pl)	organs where sperm produced, two testes in drone
thorax	middle section of bee where wings and legs attach
tomentum (s) tomenta (pl)	bands of hairs on the tergites of the abdomen
vaginal chamber, passage and orifice	opening from common oviduct to opening into sting chamber of queen
ventral	underside of bee
ventriculus	midgut or main digestive stomach where food is absorbed
waist	narrow region between thorax and abdomen
wax glands	four pairs of glands secrete liquid wax on underside of abdomen

Bibliography

Allan, R. and Greenwood, T. 2001. *Year 12 Biology: Student resource and activity manual.* Hamilton, New Zealand: Biozone.390pp.

Cobey, S.W. 1998. *Instrumental insemination of honey bee queens.* Columbus Ohio: Ohio State University. 25 minute training video.

Dade, H. A. 1985. *Anatomy and dissection of the honey bee.* London: International Bee Research Association.158pp. and 20 plates.

Dews, J.E. and Milner, E. 2004. *Breeding better bees: using simple modern methods* (3rd edition). Great Britain: British Isles Bee Breeders Association. 64pp.

Dietz, A. 1993. *In*: Graham J.M. (Ed.) 1993. *The hive and the honey bee.* Hamilton, Illinois: Dadant and Sons Inc.p36.

Gary, N.E. 1963. Observations of mating behaviour in the honey bee. *Journal of Apicultural Research* 2:3-13.

Gary, N.E. 1992. *In*: Graham J.M. (Ed.) 1993. *The hive and the honey bee.* Hamilton, Illinois: Dadant and Sons Inc.p353.

Fert, G. 1997. *Breeding queens: production of package bees, introduction to instrumental insemination* (3rd edition). Argentan, France: O.P.I.D.A. 104pp.

Foote, L. 1971. California nosema survey 1969-1970. *American Bee Journal* 111: 17.

Goodwin, M and Van Eaton, C. 1999. *Elimination of American foulbrood without the use of drugs.* Tauranga, New Zealand: National Beekeepers' Association of New Zealand (Inc). 78pp.

Goodwin, M and Van Eaton, C. 2001. *Control of varroa: a guide for New Zealand Beekeepers.* Wellington, New Zealand: Ministry of Agriculture and Forestry. 120pp.

Graham J.M. (Ed.) 1993. *The hive and the honey bee.* Hamilton, Illinois: Dadant and Sons.1324pp.

Kleinschmidt, G.J and Kondos, A.C. 1976. The influence of crude protein levels on colony production. *Australian Beekeeper Journal* 78(2): 36-39.

Laidlaw, H.H. 1977. *Instrumental insemination of honey bee queens: pictorial instructional manual.* Hamilton, Illinois: Dadant and Sons.144pp.

Laidlaw, H.H. and Page, R.E. 1997. *Queen rearing and bee breeding.* Cheshire, Connecticut: Wicwas Press. 224pp.

Matheson, A. 1985. *Beekeeping -Apiary Sites: How to reduce drifting.* MAF Aglink FPP 535. Wellington: Ministry of Agriculture and Fisheries.

Matheson, A. 1997. *Practical Beekeeping in New Zealand.* (3rd edition) Wellington: GP Publications. 145pp.

Morse, R.A. 1994a. *Rearing queen honey bees* (2nd edition). Connecticut: Wicwas Press. 128pp.

Morse, R.A. 1994b. *The new complete guide to beekeeping.* Woodstock, Vermont: The Countryman Press. 207pp.

Rhodes, J. and Somerville, D. 2003. *Successful introduction and performance of queen bees in a commercial apiary.* Short Report. Kingston, Australia: Rural Industries Research and Development Corporation. 3pp.

Ruttner, F. 1976. (Ed.) *The instrumental insemination of the queen bee.* (2nd edition) Bucarest, Romania: International Beekeeping Technology and Economy Institute of Apimondia. 123pp.

Snodgrass, R.E. 1956. *Anatomy of the honey bee.* Ithaca: Cornell University Press.

Snodgrass, R.E. and Erickson, E.H.1992. *In*: Graham J.M. (Ed.) 1993. *The hive and the honey bee.* Hamilton, Illinois: Dadant and Sons Inc.p166.

Stace, P. 1996. *Protein content and amino-acid profiles of honey bee-collected pollens.* Lismore, Australia: Bees 'n Trees consultants. 113pp.

van Toor, R.F. 1997. *Producing royal jelly: a guide for the commercial and hobbyist beekeeper.* Lincoln. 85pp.

von Frisch, K. 1955. *The dancing bees: an account of the life and senses of the honey bee.* (First American Edition) New York and London: Harcourt Brace Jovanovich. 182pp.

Winston, M.L. 1987. *The biology of the honey bee.* Cambridge, Massachusetts and London: Harvard University Press.281pp.

Woodward, D.R. 1990. *Food demand for colony development, crop preference and food availability for Bombus terrestris (L.) (Hymenoptera: Apidae).* Ph.D. thesis Zoology, Massey University, New Zealand 242pp.

Woodward, D.R. 1993. Ligurian bees. *American Bee Journal (February)* 1993: 124-125.

Woodward, D. R. 1993. The Ligurian Bee Story. *In: The Kangaroo Island Cookbook* p.1-4. by Wilson, M. Adelaide. 52pp.

Woodward, D.R. and Kezic, N. 1995. Breeding Ligurian queen bees (*Apis mellifera ligustica*) on Kangaroo Island, South Australia, p.32-37. *In: Proceedings of the International Symposium on bee breeding on the islands, Island of Vis, Croatia. April 19-26, 1995.*

Woodward, D.R. 1998. Advanced technologies in apiculture, p.34-45. *In: Proceedings of the '98 International Symposium in Apicultural Science 2-5 November 1998, Suwon, South Korea.*